I0147369

George Paxton Young

Principles of the Solution of Equations of the Higher Degrees

With Applications

George Paxton Young

Principles of the Solution of Equations of the Higher Degrees
With Applications

ISBN/EAN: 9783337157753

Printed in Europe, USA, Canada, Australia, Japan

Cover: Foto ©Thomas Meinert / pixelio.de

More available books at **www.hansebooks.com**

PRINCIPLES

OF THE

SOLUTION OF EQUATIONS OF THE HIGHER DEGREES,

WITH APPLICATIONS.

BY GEORGE PAXTON YOUNG,

Toronto, Canada.

CONTENTS.

1. Conception of a simple state to which every algebraical expression can be reduced. §6.

2. The unequal particular cognate forms of the generic expression under which a given simplified expression falls are the roots of a rational irreducible equation; and each of the unequal particular cognate forms occurs the same number of times in the series of the cognate forms. §9, 17.

3. Determination of the form which a rational function of the primitive n^{th} root of unity ω_1 and of other primitive roots of unity must have, in order that the substitution of any one of certain primitive n roots of unity, ω_1, ω_2, ω_3, etc., for ω_1 in the given function may leave the value of the function unaltered. Relation that must subsist among the roots ω_1, ω_2, etc., that satisfy such a condition. §20.

4. If a simplified expression which is the root of a rational irreducible equation of the Nth degree involve a surd of the highest rank (§3) not a root of unity, whose index is $\frac{1}{m}$, the denominator of the index being a prime number, N is a multiple of m. But if the simplified root involve no surds that are not roots of unity, and if one of the surds involved in it be the primitive n^{th} root of unity, N is a multiple of a measure of $n-1$. §28.

5. Two classes of solvable equations. §30.

6. The simplified root r_1 of a rational irreducible equation $F(x)=0$ of the mth degree, m prime, which can be solved in algebraical functions, is of the form

$$r_1 = \frac{1}{m}\left(g + \Delta_1^{\frac{1}{m}} + a_1\Delta_1^{\frac{2}{m}} + b_1\Delta_1^{\frac{3}{m}} + \ldots + e_1\Delta_1^{\frac{m-2}{m}} + h_1\Delta_1^{\frac{m-1}{m}}\right);$$

where g is rational, and a_1, b_1, etc., involve only surds subordinate to $\Delta_1^{\frac{1}{m}}$. §38, 47.

7. The equation $F(x) = 0$ has an auxiliary equation of the $(m - 1)^{\text{th}}$ degree. §35, 52.

8. If the roots of the auxiliary be $\Delta_1, \delta_2, \delta_3, \ldots, \delta_{m-1}$, the $m - 1$ expressions in each of the groups

$$\Delta_1^{\frac{1}{m}}\delta_{m-1}^{\frac{1}{m}},\quad \delta_2^{\frac{1}{m}}\delta_{m-2}^{\frac{1}{m}}, \ldots, \quad \delta_{m-1}^{\frac{1}{m}}\Delta_1^{\frac{1}{m}},$$

$$\Delta_1^{\frac{2}{m}}\delta_{m-2}^{\frac{1}{m}},\quad \delta_2^{\frac{2}{m}}\delta_{m-4}^{\frac{1}{m}}, \ldots, \quad \delta_{m-1}^{\frac{2}{m}}\delta_2^{\frac{1}{m}},$$

$$\Delta_1^{\frac{3}{m}}\delta_{m-3}^{\frac{1}{m}},\quad \delta_2^{\frac{3}{m}}\delta_{m-6}^{\frac{1}{m}}, \ldots, \quad \delta_{m-1}^{\frac{3}{m}}\delta_3^{\frac{1}{m}},$$

and so on, are the roots of a rational equation of the $(m-1)^{\text{th}}$ degree. The $\frac{m-1}{2}$ terms

$$\Delta_1^{\frac{1}{m}}\delta_{m-1}^{\frac{1}{m}},\quad \delta_2^{\frac{1}{m}}\delta_{m-2}^{\frac{1}{m}}, \ldots, \quad \delta_{\frac{m-1}{2}}^{\frac{1}{m}}\delta_{\frac{m+1}{2}}^{\frac{1}{m}},$$

are the roots of a rational equation of the $\left(\frac{m-1}{2}\right)^{\text{th}}$ degree. §39, 44, 55.

9. Wider generalization. §45, 57.

10. When the equation $F(x) = 0$ is of the first class, the auxiliary equation of the $(m - 1)^{\text{th}}$ degree is irreducible. §35. Also the roots of the auxiliary are rational functions of the primitive m^{th} root of unity. §36. And, in the particular case when the equation $F(x) = 0$ is the reducing Gaussian equation of the mth degree to the equation $x^n - 1 = 0$, each of the $\frac{m-1}{2}$ expressions,

$$\Delta_1^{\frac{1}{m}}\delta_{m-1}^{\frac{1}{m}},\quad \delta_2^{\frac{1}{m}}\delta_{m-2}^{\frac{1}{m}}, \text{ \&c.},$$

has the rational value n. §41. Numerical verification. §42.

11. Solution of the Gaussian. §43.

12. Analysis of solvable irreducible equations of the fifth degree. The auxiliary biquadratic either is irreducible, or has an irreducible sub-auxiliary of the second degree, or has all its roots rational. The three cases considered separately. Deduction of Abel's expression for the roots of a solvable quintic. §58-74.

PRINCIPLES.

§1. It will be understood that the surds appearing in the present paper have *prime numbers* for the denominators of their indices, unless where the contrary is expressly stated. Thus, $2^{\frac{1}{15}}$ may be regarded as $h^{\frac{1}{5}}$, a surd with the index $\frac{1}{5}$, h being $2^{\frac{1}{3}}$. It will be understood also that no surd appears in the denominator of a fraction. For instance, instead of $\dfrac{2}{1 + \sqrt{-3}}$ we should write $\dfrac{1 - \sqrt{-3}}{2}$. When a surd is spoken of as occurring in an algebraical expression, it may be present in more than one of its powers, and need not be present in the first.

§2. In such an expression as $\sqrt{2} + (1 + \sqrt{2})^{\frac{1}{3}}$, $\sqrt{2}$ is *subordinate* to the *principal* surd $(1 + \sqrt{2})^{\frac{1}{3}}$, the latter being the only principal surd in the expression.

§3. A surd that has no other surd subordinate to it may be said to be *of the first rank;* and the surd $h^{\frac{1}{a}}$, where h involves a surd of the $(a - 1)^{th}$ rank, but none of a higher rank, may be said to be *of the ath rank.* In estimating the rank of a surd, the denominators of the indices of the surds concerned are always supposed to be prime numbers. Thus, $3^{\frac{1}{4}}$ is a surd of the second rank.

§4. An algebraical expression in which $\Delta_1^{\frac{1}{m}}$ is a principal (see §2) surd may be arranged according to the powers of $\Delta_1^{\frac{1}{m}}$ lower than the mth, thus,

$$\frac{1}{m}\left(g_1 + k_1 \Delta_1^{\frac{1}{m}} + a_1 \Delta_1^{\frac{2}{m}} + b_1 \Delta_1^{\frac{3}{m}} + \ldots + e_1 \Delta_1^{\frac{m-2}{m}} + h_1 \Delta_1^{\frac{m-1}{m}}\right) \quad (1)$$

where g_1, k_1, a_1, etc., are clear of $\Delta_1^{\frac{1}{m}}$.

§5. If an algebrical expression r_1, arranged as in (1), be zero, while the coefficients g_1, k_1, etc., are not all zero, an equation

$$\omega \Delta_1^{\frac{1}{m}} = l_1 \qquad (2)$$

must subsist; where ω is an mth root of unity; and l_1 is an expression involving only such surds exclusive of $\Delta_1^{\frac{1}{m}}$ as occur in r_1. For, let the first of the coefficients h_1, e_1, etc., proceeding in the order of the descending powers of $\Delta_1^{\frac{1}{m}}$, that is not zero, be n_1, the coefficient of $\Delta_1^{\frac{s}{m}}$. Then we may put

$$mr_1 = n_1\left\{f\left(\Delta_1^{\frac{1}{m}}\right)\right\} = n_1 \Delta_1^{\frac{s}{m}} + \text{etc.} = 0.$$

Because $\Delta_1^{\frac{1}{m}}$ is a root of each of the equations $f(x) = 0$ and $x^m - \Delta_1 = 0$, $f(x)$ and $x^m - \Delta_1$ have a common measure. Let their H. C. M., involving only such surds as occur in $f(x)$ and $x^m - \Delta_1$, be $\varphi(x)$. Then, because $\varphi(x)$ is a measure of $x^m - \Delta_1$, the roots of the equation

$$\varphi(x) = x^c + p_1 x^{c-1} + p_2 x^{c-2} + \text{etc.} = 0$$

are $\Delta_1^{\frac{1}{m}}$, $\omega_1\Delta_1^{\frac{1}{m}}$, $\omega_2\Delta_1^{\frac{1}{m}}$, ..., $\omega_{c-1}\Delta_1^{\frac{1}{m}}$; where ω_1, ω_2, etc., are distinct primitive m^{th} roots of unity. Therefore,

$$\Delta_1^{\frac{c}{m}}(\omega_1\,\omega_2..)(-1)^c = p_c.$$

Now c is a whole number less than m but not zero; and, by §1, m is prime. Therefore there are whole numbers n and h such that

$$\Delta_1^{\frac{cn}{m}}(\omega_1\,\omega_2..)^n(-1)^{cn} = \Delta_1^{\frac{1}{m}}\Delta_1^h(\omega_1\,\omega_2..)^n(-1)^{cn} = p_c^n.$$

Therefore, if $(\omega_1\,\omega_2..)^n = \omega$, and $l_1\Delta_1^h(-1)^{cn} = p_c^n$, $\omega\Delta_1^{\frac{1}{m}} = l_1$.

§6. Let r_1 be an algebraical expression in which no root of unity having a rational value occurs in the surd form $1^{\frac{1}{m}}$. Also let there be in r_1 no surd $\Delta_1^{\frac{1}{m}}$ not a root of unity, such that

$$\Delta_1^{\frac{1}{m}} = e_1 , \tag{3}$$

where e_1 is an expression involving no surds of so high a rank as $\Delta_1^{\frac{1}{m}}$ except such as either are roots of unity, or occur in r_1 being at the same time distinct from $\Delta_1^{\frac{1}{m}}$. The expression r_1 may then be said to have been *simplified* or to be *in a simple state.*

§7. Some illustrations of the definition in §6 may be given. The root $8^{\frac{1}{3}}$ cannot occur in a simplified expression r_1; for its value is 2ω, ω being a third root of unity; but the equation $8^{\frac{1}{3}} = 2\omega$ is of the inadmissible type (3). Again, the root $\sqrt{5}$ cannot occur in a simplified expression; for, ω_1 being a primitive fifth root of unity, $\sqrt{5} = 2(\omega_1 + \omega_1^4) + 1$; an equation of the type (3). Once more, a root of the cubic equation $x^3 - 3x - 4 = 0$, in the form $(2 + \sqrt{3})^{\frac{1}{3}} + (2 - \sqrt{3})^{\frac{1}{3}}$, is not in a simple state, because $(2 - \sqrt{3})^{\frac{1}{3}} = (2 - \sqrt{3})(2 + \sqrt{3})^{\frac{2}{3}}$.

§8. Let
$$p_1\Delta_1^{\frac{m-1}{m}} + p_2\Delta_1^{\frac{m-1}{m}} + \,..\, + p_m = 0; \tag{4}$$

where $\Delta_1^{\frac{1}{m}}$ is a surd occurring in a simplified expression r_1; and p_1, p_2, etc., involve no surds of so high a rank as $\Delta_1^{\frac{1}{m}}$, except such as either are roots of unity, or occur in r_1 being at the same time distinct from $\Delta_1^{\frac{1}{m}}$. The coefficients p_1, p_2, etc., must be zero separately. For, by §5, if they were not, we should have $\omega\Delta_1^{\frac{1}{m}} = l_1$, ω being an mth root of unity, and l_1 involving only surds in (4) distinct from $\Delta_1^{\frac{1}{m}}$; an equation of the inadmissible type (3).

§9. The expression r_1 being in a simple state, we may use R as a generic symbol to include the various particular expressions, say r_1, r_2, r_3, etc., obtained by assigning all their possible values to the surds involved in r_1, with the restriction that, where the base of a surd is unity, the rational value of the surd is not to be taken into account. These particular expressions, not necessarily all unequal, may be called *the particular cognate forms of* R. For instance, if $r_1 = 1^{\frac{1}{3}}$, R has two particular cognate forms, the rational value of the third root of unity not being counted. If $r_1 = (1 + \sqrt{2})^{\frac{1}{3}}$, R has six particular cognate forms all unequal. Should $r_1 = (2 + \sqrt{3})^{\frac{1}{3}} + (2 - \sqrt{3})(2 + \sqrt{3})^{\frac{2}{3}}$, R has six particular cognate forms, but only three unequal, each of the unequal forms occurring twice.

§10. PROPOSITION I. An algebraical expression r_1 can always be brought to a simple state.

For r_1 may be cleared of all surds such as $1^{\frac{1}{m}}$ having a rational value. Suppose that r_1 then involves a surd $\Delta_1^{\frac{1}{m}}$, not a root of unity, by means of which an equation such as (3) can be formed. Substitute for $\Delta_1^{\frac{1}{m}}$ in r_1 its value e_1 as thus given. The result will be to eliminate $\Delta_1^{\frac{1}{m}}$ from r_1 without introducing into the expression any *new* surd as high in rank as $\Delta_1^{\frac{1}{m}}$, and at the same time not a root of unity. By continuing to make all the eliminations of this kind that are possible, we at last reach a point where no equation of the type (3) can any longer be formed. Then because, by the course that has been pursued, no roots of the form $1^{\frac{1}{m}}$ having a rational value have been left in r_1, r_1 is in a simple state.

§11. It is known that, if N be any whole number, the equation whose roots are the primitive N^{th} roots of unity is rational and irreducible.

§12. Let N be the continued product of the distinct prime numbers n, a, b, etc. Let ω_1 be a primitive n^{th} root of unity, θ_1 a primitive a^{th} root of unity, and so on. Let ω represent any one indifferently of the primitive n^{th} roots of unity, θ any one indifferently of the primitive a^{th} roots of unity, and so on. Let $f(\omega_1, \theta_1, \text{etc.})$ be a rational function of ω_1, θ_1, etc. Then a corollary from §11 is, that if $f(\omega_1, \theta_1, \text{etc.}) = 0$, $f(\omega, \theta, \text{etc.}) = 0$. For t_1 being a primitive N^{th} root of unity, and t representing any one indifferently of the primitive N^{th} roots of unity, we may put

$$f(\omega_1, \theta_1, \text{etc.}) = a_1 t_1^{N-1} + a_2 t_1^{N-2} + \text{etc.} = 0,$$

$$\text{and } f(\omega, \theta, \text{etc.}) = a_1 t^{N-1} + a_2 t^{N-2} + \text{etc.};$$

where the coefficients a_1, a_2, etc., are rational. Should these coefficients be all zero, $f(\omega, \theta, \text{etc.}) = 0$. Should they not be all zero, let a_r be the first that is not zero. Then we may put

$$f(\omega_1, \theta_1, \text{etc.}) = a_r \left\{ \varphi(t_1) \right\} = a_r t_1^{N-r} + \text{etc.} = 0.$$

Therefore, t_1 is a root of the rational equation $\varphi(x) = 0$, being at the same time a root of the rational (see §11) equation $\psi(x) = 0$, whose roots are the primitive N^{th} roots of unity. Hence $\psi(x)$ and $\varphi(x)$ have a common measure. But by §11, $\psi(x)$ is irreducible. Therefore it is a measure of $\varphi(x)$; and the roots of the equation $\psi(x) = 0$ are roots of the equation $\varphi(x) = 0$. Therefore,

$$f(\omega, \theta, \text{etc.}) = a_r\left\{\varphi(t)\right\} = 0.$$

§13. Another corollary is, that if

$$f(\omega_1, \theta_1, \text{etc.}) = h_1\omega_1^{n-1} + h_2\omega_1^{n-2} + \ldots + h_n = 0,$$

where h_1, h_2, etc., are clear of ω_1, the coefficients h_1, h_2, etc., are all equal to one another. For, by §12, because $f(\omega_1, \theta_1, \text{etc.}) = 0$, $f(\omega, \theta_1, \text{etc.}) = 0$. Therefore $\omega\left\{f(\omega, \theta_1, \text{etc.})\right\} = 0$. In $\omega\left\{f(\omega, \theta_1, \text{etc.})\right\}$ give ω successively its $n - 1$ different values. Then, in addition,

$nh_1 = h_1 + h_2 + \ldots + h_n$. Similarly, $nh_2 = h_1 + h_2 + \ldots + h_n \therefore h_1 = h_2$. In like manner all the terms h_1, h_2, etc., are equal to one another.

§14. Proposition II. If the simplified expression r_1, one of the particular cognate forms of R, be a root of the rational equation $F(x) = 0$, all the particular cognate forms of R are roots of that equation.

For, let r_2 be a particular cognate form of R. By §12, the law to be established holds when there are no surds in r_1 that are not roots of unity. It will be kept in view that, according to §1, when roots of unity are spoken of, such roots are meant as $1^{\frac{1}{m}}$, m being a prime number. Assume the law to have been found good for all expressions that do not involve more than $n - 1$ distinct surds that are nct roots of unity; then, making the hypothesis that r_1 involves not more than n distinct surds that are not roots of unity, the law can be shown still to hold; in which case it must hold universally. For, let $\Delta_1^{\frac{1}{m}}$, not a root of unity, be a surd of the highest rank (see §3) in r_1. Then $F(r_1)$ may be taken to be the expression (1), and $F(r_2)$ to be the expression formed from (1) by selecting particular values of the surds involved under the restriction specified in §9. In passing from r_1 to r_2, let $\Delta_1^{\frac{1}{m}}$, a_1, etc., become respectively $\Delta_2^{\frac{1}{m}}$, a_2, etc. Then

$$m\left\{F(r_1)\right\} = h_1\,\Delta_1^{\frac{m-1}{m}} + e_1\,\Delta_1^{\frac{m-2}{m}} + \text{etc.} = 0,$$

$$\text{and } m\left\{F(r_1)\right\} = h_2\,\Delta_2^{\frac{m-1}{m}} + e_2\,\Delta_2^{\frac{m-2}{m}} + \text{etc.}$$

By §8, because r_1 is in a simple state, and $F(r_1) = 0$, the coefficients h_1, e_1, etc., are zero separately. But h_1 is clear of the surd $\Delta_1^{\frac{1}{m}}$. It therefore does not involve more than $n - 1$ distinct surds that are not roots of unity. Therefore, on the assumption on which we are proceeding, because $h_1 = 0$, $h_2 = 0$. In like manner, $e_2 = 0$, and so on. Therefore $F(r_2) = 0$.

§15. *Cor.* Let the simplified expression r_1 be the root of an equation $F(x) = 0$ whose coefficients involve certain surds $z_1^{\frac{1}{n}}$, $u_1^{\frac{1}{s}}$, etc., that have the same determinate values in r_1 as in $F(x)$. Then, if r_2 be a particular cognate form of R in which the surds $z_1^{\frac{1}{n}}$, $u_1^{\frac{1}{s}}$, etc., retain the determinate values belonging to them in r_1, r_2 is a root of the equation $F(x) = 0$. For, $F(r_1) = 0$. Therefore, by the Proposition, $F(R) = 0$. Let R, restricted by the condition that the surds $z_1^{\frac{1}{n}}$, $u_1^{\frac{1}{s}}$, etc., retain the determinate values belonging to them in r_1, be R'. Then $F(R') = 0$. A particular case of this is $F(r_2) = 0$. The corollary established simply means that the surds $z_1^{\frac{1}{n}}$, $u_1^{\frac{1}{s}}$, etc., may be taken to be rational for the purpose in hand.

§16. The simplified expression r_1 being one of the particular cognate forms of R, let

$$r_1, r_a, \text{etc.} \qquad (5)$$

be the entire series of the particular cognate forms of R, not necessarily unequal to one another. Then, if the equation whose roots are the terms in (5) be $X = 0$, X is rational. In like manner, if those particular cognate forms of R, not necessarily unequal, that are obtained when certain surds $z_1^{\frac{1}{n}}$, $u_1^{\frac{1}{s}}$, etc., retain the determinate values belonging to them in r_1, be

$$r_1, r_e, \text{etc.} \qquad (6)$$

and if the equation whose roots are the terms in (6) be $X' = 0$, X' involves only surds found in the series $z_1^{\frac{1}{n}}$, $u_1^{\frac{1}{s}}$, etc. This is substantially proved by Legendre in his Théorie des Nombres, §487, third edition.

§17. PROPOSITION III. The unequal particular cognate forms of R, the generic expression under which the simplified expression r_1 falls, are the roots of a rational irreducible equation; and each of the unequal particular cognate forms occurs the same number of times in the series of the cognate forms.

As in §16, let the entire series of the particular cognate forms of R be the terms in (5), the equation that has these terms for its roots being $X = 0$. By §16, X is rational. Should X not be irreducible, it has a rational irreducible factor, say $F(x)$, such that r_1 is a root of the equation $F(x) = 0$. By Prop. II., because r_1 is in a simple state, all the terms in (5) are roots of the equation $F(x) = 0$, while at the same time, because $F(x)$ is a factor of X, all the roots of the equation are terms in (5). And the equation $F(x) = 0$, being irreducible, has no equal roots. Therefore its roots are the unequal terms in (5). Should $F(x)$ not be identical with X, put

$$X = \{F(x)\} \{\varphi(x)\}.$$

Because X and $F(x)$ are rational, $\varphi(x)$ is rational. Then, since $\varphi(x)$ is a measure of X, and the equation $F(x) = 0$ has for its its roots the unequal roots of the equation $X = 0$, the equations $F(x) = 0$ and $\varphi(x) = 0$ have a root in common. Consequently, since $F(x)$ is irreducible, it is a measure of $\varphi(x)$. Therefore $\{F(x)\}^2$ is a measure of X. Going on in this way we ultimately get $X = \{F(x)\}^N$; which means that each of the particular cognate forms of R has its value repeated N times in the series of the particular cognate forms.

§18. *Cor.* 1. The series (6) consisting of those particular cognate forms of R in which certain surds $z_1^{\frac{1}{n}}, u_1^{\frac{1}{s}}$, etc., retain the determinate values belonging to them in r_1, each of the unequal terms in (6) occurs the same number of times in (6); and the unequal terms in (6) are the roots of an irreducible equation whose coefficients involve only surds found in the series $z_1^{\frac{1}{n}}, u_1^{\frac{1}{s}}$, etc. Should X' not be irreducible, by which in such a case is meant incapable of being broken into lower factors involving only surds occurring in X', let it have the irreducible factor X''. That is to say, X'' involves only surds occurring in X', and has itself no lower factor involving only surds that occur in X''. We may take r_1 to be a root of the equation $X'' = 0$. Then, by Cor. Prop. II., all the terms in (6) are roots of that equation, all the roots of the equation being at the same time terms in (6). And the equation $X'' = 0$ being irreducible, has no equal roots. Therefore its roots are the unequal terms in (6). Put

2

$X' = (X'')(X''')$. Then, by the line of reasoning followed in the Proposition, X''' has a measure identical with X''. And so on. Ultimately $X' = (X'')^N$.

§19. *Cor.* 2. If r_2, one of the particular cognate forms of R, be zero, all the particular cognate forms of R are zero. For, by the proposition, the particular cognate forms of R are the roots of a rational irreducible equation $F(x) = 0$. And r_2, one of the roots of that equation, is zero, but the only rational irreducible equation that has zero for a root is $x = 0$. Therefore $F(x) = x = 0$. In fact, in the case supposed, the simplified expression r_1 is zero, and R has no particular cognate forms distinct from r_1.

§20. Proposition IV. Let N be the continued product of the distinct prime numbers n, a, etc. Let ω_1 be a primitive n^{th} root of unity, θ_1 a primitive a^{th} root of unity, and so on. Then if the equation

$$F(x) = x^d + b_1x^{d-1} + b_2x^{d-2} + \text{etc.} = 0$$

be one in which the coefficients b_1, b_2, etc., are rational functions of ω_1, θ_1, etc., and if all the primitive n^{th} roots of unity, which, when substituted for ω_1 in $F(x)$, leave $F(x)$ unaltered, be

$$\omega_1\,,\ \omega_2\,,\ \ldots.,\ \omega_s\,, \tag{7}$$

the series (7) either consists of a single term or it is made up of a cycle of primitive n^{th} roots of unity,

$$\omega_1\,,\ \omega_1^{\lambda}\,,\ \omega_1^{\lambda^2}\,,\ \ldots.,\ \omega_1^{\lambda^{s-1}}\,; \tag{18}$$

that is to say, no term in (8) after the first is equal to the first, but $\omega_1^{\lambda^s} = \omega_1$. Also, if (let it be kept in view that n is prime) the cycle that contains all the primitive n^{th} roots of unity be

$$\omega_1,\ \omega_1^{\beta}\,,\ \omega_1^{\beta^2}\,,\ \ldots.,\ \omega_1^{\beta^{n-2}}\,, \tag{9}$$

and if C_1 be the sum of the terms in the cycle (8), the form of $F(x)$ is

$$F(x) = x^d - (p_1C_1 + p_2C_2 + \ \ldots. + p_mC_m)\,x^{d-1} + (q_1C_1 + q_2C_2 + \text{etc.})\,x^{d-2} + \text{etc.} \tag{10}$$

where each of the expressions in the series C_1, C_2, C_3, etc., is what the immediately preceding term becomes by changing ω_1 into ω_1^{β}, C_m through this change becoming C_1; and p_1, p_2, q_1, etc., are clear of ω_1.

For, assuming that there is a term ω_2 in (7) additional to ω_1, we may take ω_2 to be the first term in (9) after ω_1 that occurs in (7); and it may be considered to be $\omega_1^{\beta^m}$, which may be otherwise written ω_1^{λ}. Then, if $F(x)$ be written $\varphi(\omega_1)$, we have by hypothesis

$\varphi(\omega_1) = \varphi(\omega_1^{\lambda})$. Therefore, by §12, changing ω_1 into ω_1^{λ}, $\varphi(\omega_1^{\lambda}) = \varphi(\omega_1^{\lambda^2})$. Therefore $\varphi(\omega_1) = \varphi(\omega_1^{\lambda^2})$. And thus ultimately $\varphi(\omega_1) = \varphi(\omega_1^{\lambda^z})$, or $\varphi(\omega_1) = \varphi(\omega_1^{\beta^{mz}})$, z being any whole number positive or negative. But $\omega_1^{\lambda^z}$ includes all the terms in (8). Therefore each of these terms is a term in (7). Suppose if possible that there is a term in (7), say $\omega_1^{\beta^h}$, which does not occur in (8). Then, just as we deduced $\varphi(\omega_1) = \varphi(\omega_1^{\beta^{mz}})$ from the equation $\varphi(\omega_1) = \varphi(\omega_1^{\beta^m})$, we can, because still farther $\varphi(\omega_1) = \varphi(\omega_1^{\beta^h})$, deduce $\varphi(\omega_1) = \varphi(\omega_1^{\beta^{mz+hu}})$. Because $\omega_1^{\beta^h}$ lies outside the cycle (8), h is not a multiple of m. And it is not less than m, because $\omega_1^{\beta^m}$ is the first term in (9) after ω_1, which, when substituted for ω_1 in $\varphi(\omega_1)$, leaves $\varphi(\omega_1)$ unaltered. Therefore $h = qm + v$, where q and v are whole numbers, and v is less than m but not zero. Put

$$z = -(h+q), \text{ and } u = m+1 \therefore mz+hu = v \therefore \varphi(\omega_1) = \varphi(\omega_1^{\beta^v});$$

which, because v is less than m but not zero, and $\omega_1^{\beta^m}$ is the first term in (9) after ω_1 which, when substituted for ω_1 in $\varphi(\omega_1)$, leaves $\varphi(\omega_1)$ unaltered, is impossible. Hence, no term in (7) lies outside the cycle (8), while it has also been shown that all the terms in (8) are terms in (7). Therefore the terms in (7) are identical with those constituting the cycle (8). We have now to determine the form of $F(x)$. The expressions, C_1, C_2, etc., taken together, are the sum of the terms in (9). Therefore $C_1 + C_2 + \dots + C_m = -1$. (11)
Because (9) contains all the primitive n^{th} roots of unity, we may put

$$F(x) = x^d - \{p + (p+p_1)\,\omega_1 + (p+p_2)\,\omega_1^{\beta} + \text{etc.}\}\,x^{d-1} + \text{etc.}; \quad (12)$$

where p, p_1, etc., are clear of ω_1. But $F(x)$ remains unaltered when ω_1 is changed into $\omega_1^{\beta^m}$. Therefore

$$F(x) = x^d - \{p + (p+p_1)\,\omega_1^{\beta^m} + \text{etc.}\}\,x^{d-1} + \text{etc.} \quad (13)$$

Therefore, equating the coefficients of x^{d-1} in (12) and (13),

$$(p - p_1) + \dots + (p_{m+1} - p_1)\,\omega_1^{\beta^m} + \text{etc.} = 0.$$

Here, by §13, the coefficients of the different powers of ω_1 have all the same value. And one of them, $p - p_1$, is zero Therefore

$p_{m+1} = p_1$. That is to say, the coefficient of $\omega_1^{\beta^m}$ or ω_1^{λ} is the same as that of ω_1. In like manner the coefficients of all the terms in (8) are the same. Therefore one group of the terms that together make up the coefficient of x^{d-1} in (12) is properly represented by $-(p+p_1)C_1$. In the same way another group is properly represented by $-(p+p_2)C_2$, and so on. Hence

$$F(x) = x^d - \{p + (p+p_1)\,C_1 + (p+p_2)\,C_2 + \text{etc.}\}x^{d-1} + \text{etc.}$$

And by (11) this is equivalent to (10). The form of $F(x)$ has been deduced on the assumption that the series (7) contains more than one term; but, should the series (7) consist of a single term, the result obtained would still hold good, only in that case each of the expressions C_1, C_2, etc., would be a primitive n^{th} root of unity.

§21. A simplified expression will not cease to be in a simple state, if we suppose that any surd that can be eliminated from it, without the introduction of any new surd, has been eliminated.

§22. Proposition V. In the simplified expression r_1, one of the particular cognate forms of R, modified according to §21, let the surd $\Delta_1^{\frac{1}{m}}$ of the highest rank be not a root (see §1) of unity. Then, if the particular cognate forms of R obtained by changing $\Delta_1^{\frac{1}{m}}$ in r_1 successively into the different m^{th} roots of the determinate base Δ_1, be

$$r_1, r_2, \ldots, r_m, \tag{14}$$

these terms are all unequal.

For the terms in (14) are all the particular cognate forms of R obtained when we allow all the surds in r_1 except $\Delta_1^{\frac{1}{m}}$ to retain the determinate values belonging to them in r_1. Therefore, by Cor. 1, Prop. III., each of the unequal terms in (14) has its value repeated the same number of times in that series. Let u be the number of the unequal terms in (14), and let each occur c times. Then $uc = m$. Suppose if possible that $u = 1$. This means that all the terms in (14) are equal. Therefore, r_1 being the expression (1),

$$mr_1 = r_1 + r_2 + \ldots + \text{etc.} = g_1.$$

Therefore the surd $\Delta_1^{\frac{1}{m}}$ can be eliminated from r_1 without the introduction of any new surd; which, by §21, is impossible. Therefore u is not unity. But, by §1, m is a prime number. And $m = uc$. Therefore $c = 1$ and $u = m$. This means that all the terms in (14) are unequal.

§23. *Cor.* 1. Let r_{a+1} be any one of the particular cognate forms of R; and let $\Delta_{a+1}^{\frac{1}{m}}$, h_{a+1}, etc., be respectively what $\Delta_1^{\frac{1}{m}}$, h_1, etc., become in passing from r_1 to r_{a+1}. Also let the m particular cognate forms of R, obtained by changing $\Delta_{a+1}^{\frac{1}{m}}$ in r_{a+1} successively into the different m^{th} roots of Δ_{a+1}, be

$$r_{a+1}, \; r_{a+2}, \; \ldots\ldots, \; r_{a+m}. \tag{15}$$

These terms are all unequal. For, because $\Delta_1^{\frac{1}{m}}$ is a principal surd in r_1, and r_2 is what r_1 becomes when $\Delta_1^{\frac{1}{m}}$ is changed into a surd whose value is $\omega_1 \Delta_1^{\frac{1}{m}}$, ω_1 being a primitive m^{th} root of unity. the view may be taken that r_2 involves no surds additional to those found in r_1, except the primitive m^{th} root of unity ω_1. Therefore $r_1 - r_2$ involves no surds distinct from primitive m^{th} roots of unity that are not found in the simplified expression r_1. Therefore $r_1 - r_2$ is in a simple state. Let r_{a+2} be what r_{a+1} becomes by changing $\Delta_{a+1}^{\frac{1}{m}}$ into $\omega_1 \Delta_{a+1}^{\frac{1}{m}}$. Then $r_{a+1} - r_{a+2}$ is a particular cognate form of the generic expression under which the simplified expression $r_1 - r_2$ falls. Therefore $r_{a+1} - r_{a+2}$ cannot be zero; for, if it were, $r_1 - r_2$ would, by Cor. 2, Prop. II., be zero; which, by the proposition, is impossible. Hence, the first two terms in (15) are unequal. In like manner all the terms in (15) are unequal.

§24. *Cor.* 2. Let $X_1 = 0$ be the equation whose roots are the terms in (14). When X_1 is modified according to §21, it is, by §16, clear of the surd $\Delta_1^{\frac{1}{m}}$. Should it involve any surds that are not roots of unity, take $z_1^{\frac{1}{c}}$ a surd of the highest rank not a root of unity in X_1; and, when $z_1^{\frac{1}{c}}$ is changed successively into the different c^{th} roots of the determinate base z_1, let

$$X_1, \; X_1', \; X_1'', \; \ldots\ldots, \; X_1^{(c-1)}, \tag{16}$$

be respectively what X_1 becomes. Any term in (16), as X_1', being selected, the m roots of the equation $X_1' = 0$ are unequal particular cognate forms of R. For, $z_2^{\frac{1}{c}}$ being a cth root of z_1 distinct from $z_1^{\frac{1}{c}}$, let r_{a+1} be what r_1 becomes when $z_1^{\frac{1}{c}}$ becomes $z_2^{\frac{1}{c}}$; the expressions $\Delta_1^{\frac{1}{m}}$, h_1, etc., at the same time becoming $\Delta_{a+1}^{\frac{1}{m}}$, h_{a+1}, etc. Then we may put

$$X_1 = x^m + (bz_1^{\frac{c-1}{c}} + dz_1^{\frac{c-2}{c}} + \text{etc.})\, x^{m-1} + \text{etc.}; \qquad (17)$$

where b, d, etc., are clear of $z_1^{\frac{1}{c}}$. Therefore, because r_1 is a root of the equation $X_1 = 0$,

$$\left\{ \frac{1}{m} (h_1 \Delta_1^{\frac{m-1}{m}} + \text{etc.}) \right\}^m$$

$$+ (bz_1^{\frac{c-1}{c}} + dz_1^{\frac{c-2}{c}} + \text{etc.}) \left\{ \frac{1}{m} (h_1 \Delta_1^{\frac{m-1}{m}} + \text{etc.}) \right\}^{m-1} + \text{etc.} = 0.$$

All the surds in this equation occur in the simplified expression r_1. Therefore, by Prop. II.,

$$\left\{ \frac{1}{m} (h^{a+1} \Delta_{a+1}^{\frac{m-1}{m}} + \text{etc.}) \right\}^m$$

$$+ (bz_2^{\frac{c-1}{c}} + dz_2^{\frac{c-2}{c}} + \text{etc}) \left\{ \frac{1}{m} (h_{a+1} \Delta_{a+1}^{\frac{m-1}{m}} + \text{etc.}) \right\}^{m-1} + \text{etc.} = 0.$$

Therefore $\frac{1}{m} (h_{a+1} \Delta_{a+1}^{\frac{m-1}{m}} + \text{etc.})$ or r_{a+1} is a root of the equation

$$X_1' = x^m + (bz_2^{\frac{c-1}{c}} + \text{etc.})\, x^{m-1} + \text{etc.} = 0. \qquad (18)$$

Therefore also, by Cor. Prop. II., all the terms in (15) are roots of that equation. And, by Cor. 1, the terms in (15) are all unequal. Therefore the equation $X_1' = 0$ has m unequal particular cognate forms of R for its roots.

§25. *Cor.* 3. No two of the expressions in (16), as x_1 and X_1', are identical with one another. For, in order that X_1 and X_1' might be identical, the coefficients of the several powers of x in X_1 would need to be equal to those of the corresponding powers of x in X_1'; but, if one of the coefficients of X_1 be selected in which $z_1^{\frac{1}{c}}$ is present, this coefficient can be shown to be unequal to the corresponding coefficient in X_1' in the same way in which the terms in (15) were proved to be all unequal.

§26. *Cor.* 4. Any two of the terms in (16), as X_1 and X_1', being selected, the equations $X_1 = 0$ and $X_1' = 0$ have no root in common. For, suppose, if possible, that these equations have a root in common. Taking the forms of X_1 and X_1' in (17) and (18), since r_1 is a root of the equation $X_1' = 0$,

$$r_1^m + (bz_2^{\frac{c-1}{c}} + \text{etc.})\, r_1^{m-1} + \text{etc.} = 0. \tag{19}$$

All the surds in this equation except $z_2^{\frac{1}{c}}$ occur in r_1. It is impossible that $z_2^{\frac{1}{c}}$ can occur in r_1; for, $z_1^{\frac{1}{c}}$ occurs in r_1; and $z_2^{\frac{1}{c}} = \theta_1 z_1^{\frac{1}{c}}$, θ_1 being a primitive c^{th} root of unity; but this equation, if both $z_1^{\frac{1}{c}}$ and $z_2^{\frac{1}{c}}$ occurred in r_1, would be of the inadmissible type (3). Since $z_2^{\frac{1}{c}}$ does not occur in r_1, it is a principal (see §2) surd in (19). We may, therefore, keeping in view that r_1 is the expression (1) in which $\Delta_1^{\frac{1}{m}}$ is a principal surd, arrange (19) thus,

$$\varphi\left(\Delta_1^{\frac{1}{m}}\right) = \Delta_1^{\frac{m-1}{m}}\left(p_1 z_2^{\frac{c-1}{c}} + p_2 z_2^{\frac{c-2}{c}} + \text{etc.}\right)$$
$$+\ \Delta_1^{\frac{m-2}{m}}\left(q_1 z_2^{\frac{c-1}{c}} + q_2 z_2^{\frac{c-2}{c}} + \text{etc.}\right) + \text{etc.} = 0; \tag{20}$$

where p_1, q_1, etc., are clear of $z_2^{\frac{1}{c}}$. Then, ω_1 being a primitive m^{th} root of unity such that, by changing $\Delta_1^{\frac{1}{m}}$ into the mth root of Δ_1 whose value is $\omega_1 \Delta_1^{\frac{1}{m}}$, r_1 becomes r_2,

$$\varphi\left(\omega_1\Delta_1^{\frac{1}{m}}\right) = \omega_1^{m-1}\Delta_1^{\frac{m-1}{m}}\left(p_1 z_2^{\frac{c-1}{c}} + \text{etc.}\right)$$

$$+\ \omega_1^{m-2}\Delta_1^{\frac{m-1}{m}}\left(q_1 z_2^{\frac{c-1}{c}} + \text{etc.}\right) + \text{etc.} \qquad (21)$$

The coefficients of the several powers of $\Delta_1^{\frac{1}{m}}$ in $\varphi\left(\Delta_1^{\frac{1}{m}}\right)$ cannot be all zero; for, if they were, we should have, from (21), $\varphi\left(\omega_1\Delta_1^{\frac{1}{m}}\right) = 0$. This means that r_2 is a root of the equation $X_1' = 0$. But in like manner all the terms in (14) would be roots of that equation, and X_1' would be identical with X; which, by Cor. 3, is impossible. Since the coefficients of the different powers of $\Delta_1^{\frac{1}{m}}$ in $\varphi\left(\Delta_1^{\frac{1}{m}}\right)$ are not all zero, the equation (20) gives us, by §5, $\omega\Delta_1^{\frac{1}{m}} = l_1$, ω being an mth root of unity, and l_1 involving only surds in $\varphi\left(\Delta_1^{\frac{1}{m}}\right)$exclusive of $\Delta_1^{\frac{1}{m}}$. In l_1 we may conceive $z_2^{\frac{1}{c}}$ changed into $\theta_1 z_1^{\frac{1}{c}}$. Then l_r involves only surds distinct from $\Delta_1^{\frac{1}{m}}$, all of them except the primitive c^{th} root of unity θ_1 being surds that occur in r_1. This makes the equation $\omega\Delta_1^{\frac{1}{m}} = l_1$ of the inadmissible type (3). Hence the equations $X_1 = 0$ and $X_1' = 0$ have no root in common.

§27. *Cor.* 5. Let X_2 be the continued product of the terms in (16). Then X_2, modified according to §21, is clear of $z_1^{\frac{1}{c}}$, in the same way in which X_1 is clear of $\Delta_1^{\frac{1}{m}}$. Also since, by Cor. 2, each of the equations $X_1 = 0$, $X_1' = 0$, etc., has m unequal particular cognate forms of R for its roots, and since, by Cor. 4, no two of these equations have a root in common, the mc roots of the equation $X_2 = 0$ are unequal particular cognate forms of R.

§28. PROPOSITION VI. Let the simplified expression r_1, modified according to §21, be a root of the rational irreducible equation of the Nth degree, $F(x) = 0$. Then if $\Delta_1^{\frac{1}{m}}$, not a root of unity, be a surd of the highest rank in r_1, N is a multiple of m. But if r_1 involve only surds that are roots of unity, one of them being the primitive n^{th} root of unity, N is a multiple of a measure of $n - 1$.

First, let $\Delta_1^{\frac{1}{m}}$, not a root of unity, be a surd of the highest rank in r_1. Taking the expression (1) to be r_1, let X_1 be formed as in §24, and let it be modified according to §21. It is clear of the surd $\Delta_1^{\frac{1}{m}}$. Should it involve a surd that is not a root of unity, let X_2 be formed as in §27. Setting out from r_1 we arrived by one step at X_1, an expression clear of $\Delta_1^{\frac{1}{m}}$, and such that the roots of the equation $X_1 = 0$ are unequal particular cognate forms of R. A second step brought us to X_2, an expression clear of the additional surd $z_1^{\frac{1}{c}}$, and such that the mc roots of the equation $X_2 = 0$ are unequal particular cognate forms of R. Thus we can go on till, in the series X_1, X_2, etc., we reach a term X_e into which no surds enter that are not roots of unity, the $mc \ldots\ldots l$ roots of the equation $X_e = 0$ being unequal particular cognate forms of R. Should X_e modified according to §21, not be rational, its form, by Prop. IV., putting d for $mc \ldots\ldots l$, is

$$X_e = x_d - (p_1C_1 + \ldots\ldots + p_mC_m)x^{d-1} + (q_1C_1 + \ldots\ldots + q_mC_m)x^{d-2} + \text{etc.};$$

where, one of the roots occurring in X_e being the primitive n^{th} root of unity ω_1, the coefficients p_1, q_1, etc., are clear of ω_1; and C_1 is the sum of the cycle of primitive n^{th} roots of unity (8) containing s or $\dfrac{n-1}{m}$ terms; and, the cycle (9) containing all the primitive n^{th} roots of unity, the change of ω_1 into ω_1^{β} causes C_1 to become C_2, and C_2 to become C_3, and so on, C_m becoming C_1. As was explained at the close of §20, the cycle (8) may be reduced to a single term, which is then identical with C_1. It will also not bo forgotten that the roots of unity such as the nth here spoken of are, according to §1, subject to the condition that the numbers such as n are prime. When C_1 in X_e is changed successively into C_1, C_2, etc., let X_e become

$$X_e,\ X_e',\ X_e'',\ \ldots\ldots,\ X_e^{(m-1)} \qquad (22)$$

If X_{e+1} be the continued product of the terms in (22), the dm roots of the equation $X_{e+1} = 0$ can be shown to be unequal particular cognate forms of R. For, no two terms in (22) as X_e and X'_e are identical; because, if they were, X_e would remain unaltered by the change of ω_1 into ω_1^{β}; which, by Prop. IV., because ω_1^{β} is not a term in the cycle (8), is impossible. It follows that no two of the equations $X_e = 0$, $X'_e = 0$, etc., have a root in common. For, if the equations $X_e = 0$, and $X'_e = 0$ had a root in common, since X_e and X'_e are not identical, X_e would have a lower measure involving only surds found in X'_e, because the surds in X_e are the same with those in X'_e. Let $\varphi(x)$ be this lower measure of X_e, and let r_1 be a root of the equation $\varphi(x) = 0$. Then, by Cor. Prop. II., all the d roots of the equation $X'_e = 0$ are roots of the equation $\varphi(x) = 0$; which is impossible. In the same way it can be proved that no equation in the series $X_e = 0$, $X'_e = 0$, etc., has equal roots. Since no one of these equations has equal roots, and no two of them have a root in common, the dm roots of the equation $X_{e+1} = 0$ are unequal particular cognate forms of R. Also X_{e+1}, modified according to §21, is clear of the primitive n^{th} roots of unity. Should X_{e+1} not be rational, we can deal with it as we did with X_e. Going on in this way, we ultimately reach a *rational* expression X_z such that the $dm \ldots\ldots g$ roots of the equation $X_z = 0$ are unequal particular cognate forms of R. This equation must be identical with the equation $F(x) = 0$ of which r_1 is a root. For, by Prop. III., the equation $F(x) = 0$ has for its roots the unequal particular cognate forms of R. Therefore, because the roots of the equation $X_z = 0$ are all unequal and are at the same time particular cognate forms of R, X_z must be either a lower measure of $F(x)$ or identical with $F(x)$. But $F(x)$, being irreducible, has no lower measure. Therefore X_z is identical with $F(x)$. Therefore, the equation $F(x) = 0$ being the Nth degree, $N = mc \ldots\ldots lm \ldots\ldots g$. Hence N is a multiple of m. This is the result arrived at when r_1 involves a surd of the highest rank $\Delta_1^{\frac{1}{m}}$ not a root of unity. Should r_1 involve no surds except roots (see §1) of unity, we should then have set out from X_e regarded as identical with $x - r_1$. The result would have been $N = m \ldots\ldots g$. Therefore N is a multiple of m; and, because m is here the number of cycles of s terms each, that make up the series of the primitive n^{th} roots of unity, $ms = n - 1$. Therefore N is a multiple of a measure of $n - 1$.

§29. *Cor.* Let N be a prime number. Then, if r_1 involve a surd of the highest rank $\Delta_1^{\frac{1}{m}}$ not a root (see §1) of unity, $N = m$; for, the series of integers m, c, etc., of which N is the continued product, is reduced to its first term. If r_1 involve only surds that are roots of unity, $n - 1$ is a multiple of N; for $N = m \ldots . g$; therefore, because N is prime, it is equal to m; but $ms = n - 1$; therefore $n - 1 = sN$.

THE SOLVABLE IRREDUCIBLE EQUATION OF THE mth DEGREE, m PRIME.

§30. The principles that have been established may be illustrated by an examination of the solvable irreducible rational equation of the mth degree $F(x) = 0, m$ being prime. Two cases may be distinguished, though it will be found that the roots can in the two cases be brought under a common form; the one case being that in which the simplified root r_1 is, and the other that in which it is not, a rational function of roots of unity, that is, according to §1, of roots of unity having the denominators of their indices prime numbers. The equation $F(x) = 0$ may be said to be in the former case *of the first class*, and in the latter *of the second class*.

THE EQUATION $F(x) = 0$ OF THE FIRST CLASS.

§31. In this case, by Cor. Prop. VI., r_1 being modified according to §21, if one of the roots involved in r_1 be the primitive n^{th} root of unity ω_1, $n - 1$ is a multiple of m. Also the expression written X_s in Prop. VI. is reduced to $x - r_1$, so that

$$r_1 = p_1C_1 + p_2C_2 + \ldots . + p_mC_m .$$

The m roots of the equation $F(x) = 0$ being r_1, r_2, etc, we must have

$$\left.\begin{aligned} r_1 &= p_1\,C_1 + p_2C_2 + \ldots . + p_mC_m , \\ r_2 &= p_mC_1 + p_1C_2 + \ldots . + p_{m-1}C_m , \\ &\ldots\ldots\ldots\ldots\ldots\ldots\ldots\ldots\ldots\ldots \\ r_m &= p_2\,C_1 + p_3C_2 + \ldots . + p_1C_m . \end{aligned}\right\} \qquad (23)$$

For, by Prop. II., because r_1 is a root of the equation $F(x) = 0$, all the expressions on the right of the equations (23) are roots of that equation. And no two of these expressions are equal to one another. For, take the first two. If these were equal, we should have $(p_m - p_1)\,C_1 + (p_1 - p_2)\,C_2 +$ etc. $= 0$. Therefore, by §13, each of the terms $p_m - p_1$, $p_1 - p_2$, etc., is zero. This makes p_1, p_2, etc., all equal to one another. Therefore $r_1 = -p_1$; so that the primitive n^{th} root of unity is eliminated from r_1; which, by §21, is impossible. Hence the values of the m roots of the equation $F(x) = 0$ are those given in (23).

§32. Let r_1' be one of the particular cognate forms of the generic expression R under which the simplified expression r_1 falls. Then, because, by Prop. II., all the particular cognate forms of R are roots of the equation $F(x) = 0$, r_1' is equal to one of the m terms r_1, r_2, etc., say to r_z. I will now show that the changes of the surds involved that cause r_1 to become r_1', whose value is r_z, cause r_2 to receive the value r_{z+1}, and r_3 to receive the value r_{z+2}, and so on. This may appear obvious on the face of the equations (23); but, to prevent misunderstanding, the steps of the deduction are given. Any changes made in r_1 must transform C_1 into C_z, one of the m terms C_1, C_2, etc. In passing from r_1 to r_1', while C_1 becomes C_z, let r_2 become r_2', and p_1 become p_1', and p_2 become p_2', and so on. The change that causes C_1 to become C_z transforms C_2 into C_{z+1}, and C_3 into C_{z+2}, and so on. Therefore, it being understood that p_{m+1}, C_{m+1}, etc., are the same as p_1, C_1, etc., respectively,

$$r_1' = p_1'C_z + p_2'C_{z+1} + \text{etc.},$$

$$\text{and } r_2' = p_m'C_z + p_1'C_{z+1} + \text{etc.};$$

which may be otherwise written

$$\left.\begin{aligned} r_1' &= p'_{m+2-z}\,C_1 + p'_{m+3-z}\,C_2 + \text{etc.}, \\ r_2' &= p'_{m+1-z}\,C_1 + p'_{m+2-z}\,C_2 + \text{etc.} \end{aligned}\right\} \quad (24)$$

Therefore, form (24) and (23),

$$C_1(p_{m+2-z} - p'_{m+2-z}) + C_2\,(p_{m+3-z} - p'_{m+3-z}) + \text{etc.} = 0.$$

Therefore, by §13, $p'_{m+2-z} = p_{m+2-z}$, $p'_{m+3-z} = p_{m+3-z}$, etc. Hence the second of the equations (24) becomes

$$r_2' = p_{m+1-z}\,C_1 + p_{m+2-z}\,C_2 + \text{etc.} = r_{z+1}.$$

Thus r_2 is transformed into r_{z+1}. In like manner r_3 receives the value r_{z+2}, and so on.

§33. By Cor. Prop. VI., the primitive n^{th} root of unity being one of those involved in r_1, $n - 1$ is a multiple of m. In like manner, if the primitive a^{th} root of unity be involved in r_1, $a - 1$ is a multiple of m, and so on. Therefore, if t_1 be the primitive m^{th} root of unity, t_1 is distinct from all the roots involved in r_1.

§34. From this it follows that, if the circle of roots r_1, r_2, ..., r_m, be arranged, beginning with r_c, in the order r_c, r_{c+1}, r_{c+2}, etc., and again, beginning with r_s, in the order r_s, r_{s+1}, r_{s+2}, etc., and if, t_1^a being one of the primitive m^{th} roots of unity,

$$r_c + r_{c+1} t_1 + r_{c+2} t_1^2 + \text{etc.} = r_s + r_{s+1} t_1^a + r_{s+2} t_1^{2a} + \text{etc.} \quad (25)$$

$r_c = r_s$. It is understood that in the series r_c, r_{c+1}, etc., when r_m is reached, the next in order is r_1, so that r_{m+1} is the same as r_1, and so on. In like manner r_{s+1} is the same as r_1, and so on. Since r_1, r_2, etc., do not involve the primitive m^{th} root of unity t_1, we can, by §12, substitute for t_1 in (25) successively the different primitive m^{th} roots of unity. Let this be done. Then, by addition,

$mr_c - (r_1 + r_2 + \text{etc.}) = mr_s - (r_1 + r_2 + \text{etc.})$. Therefore $r_c = r_s$.

§35. Proposition VII. Putting

$$\left.\begin{aligned} \Delta_1^{\frac{1}{m}} &= r_1 + t_1 r_2 + t_1^2 r_3 + \ldots + t_1^{m-1} r_m, \\ \Delta_2^{\frac{1}{m}} &= r_1 + t_1^2 r_2 + t_1^4 r_3 + \ldots\ldots + t_1^{2(m-1)} r_m, \\ &\ldots\ldots\ldots\ldots\ldots\ldots\ldots\ldots\ldots\ldots \\ \Delta_{m-1}^{\frac{1}{m}} &= r_1 + t_1^{-1} r_2 + t_1^{-2} r_3 + \ldots\ldots + t_1 r_m, \end{aligned}\right\} \quad (26)$$

the terms, $\qquad \Delta_1, \Delta_2, \Delta_3, \ldots\ldots, \Delta_{m-1}, \qquad (27)$

are the roots of a rational irreducible equation of the $(m-1)^{\text{th}}$ degree $\varphi(x) = 0$, which may be said to be *auxiliary* to the equation $F(x) = 0$.

For, let Δ be the generic expression of which Δ_1 is a particular cognate form; and let Δ' denote any one indifferently of the $m-1$ particular cognate forms of Δ in (27). Because, by §33, the primitive m^{th} root of unity does not enter into r_1, r_2, etc., no changes made in r_1, r_2, etc., affect t_1. Also, by §32, if r_1 becomes r_z, r_2 becomes r_{z+1}, r_3 becomes r_{z+2}, and so on. Therefore the expression

$$(r_z + tr_{z+1} + t^2 r_{z+2} + \text{etc.})^m,$$

contains all the particular cognate forms of Δ; where z may be any number in the series $1, 2, \ldots\ldots, m-1$; and t denotes any one indifferently of the primitive m^{th} roots of unity. But this is equal to

$$\{t^{1-z}(r_1 + tr_2 + t^2 r_3 + \text{etc.})\}^m \text{ or } \Delta'.$$

The conclusion established means that all the differences of value that can present themselves in the particular cognate forms of Δ must arise from the different values of t that are taken in Δ', while the expressions r_1, r_2, etc., remain unaltered. And t has not more than $m - 1$ values. Hence there are not more than $m - 1$ unequal particular cognate forms of Δ. But the $m - 1$ forms obtained by taking the different values of t in Δ' are all unequal. For, selecting t_1 and t_1^a, two distinct values of t, suppose if possible that

$$(r_1 + t_1 r_2 + \text{etc.})^m = (r_1 + t_1^a r_2 + \text{etc.})^m$$

$$\therefore t_1^s (r_1 + t_1 r_2 + \text{etc.}) = r_1 + t_1^a r_2 + \text{etc.},$$

s being a whole number. This may be written

$$r_{m+1-s} + r_{m+2-s}\, t_1 + \text{etc.} = r_1 + t_1^a r_2 + \text{etc.} \qquad (28)$$

Therefore, by §34, $r_{m+1-s} = r_1$. This means, since all the m terms r_1, r_2, etc., are unequal, that $s = 0$. Hence (28) becomes

$$r_1 + r_2 t_1 + \text{etc.} = r_1 + r_2 t_1^a + \text{etc.}$$

Therefore

$$r_2 + r_3 t_1^a + \text{etc.} = r_2 t_1^{1-a} + r_3 t_1^{2-a} + \text{etc.}$$

$$= r_{a+1} + r_{a+2} t_1 + \text{etc.}$$

Therefore, by §35, $r_2 = r_{a+1}$. Therefore, because all the m terms r_1, r_2, etc., are unequal, $a = 1$; which, because t_1 and t_1^a were supposed to be distinct primitive m^{th} roots of unity, is impossible. Therefore no two of the terms in (27) are equal to one another. And it has been proved that there is no particular cognate form of Δ which is not equal to a term in (27). Therefore the terms in (27) are the unequal particular cognate forms of Δ. Therefore, by Prop. III., they are the roots of a rational irreducible equation.

§36. PROPOSITION VIII. The roots of the equation $\varphi(x) = 0$ auxiliary (see §35) to $F(x) = 0$ are rational functions of the primitive m^{th} root of unity.

For, let the value of Δ_1, obtained from (26), and modified according to §21, be

$$\Delta_1 = k_1 + k_2 t_1 + k_3 t_1^2 + \ldots . + k_m t_1^{m-1},$$

where k_1, k_2, etc., are clear of t_1. Suppose if possible that k_1, k_2, etc., are not rational. We may take the primitive n^{th} root of unity ω_1 to be present in these coefficients. But ω_1 occurs in r_1, r_2, etc., and therefore also in Δ_1, only in the expressions C_1, C_2, etc. Therefore $\Delta_1 = d_1 C_1 + \ldots . + d_m C_m$; where d_1, etc., are clear of ω_1. The coefficients d_1, d_2, etc., cannot all be equal; for this would make $\Delta_1 = - d_1$; which, by §21, is impossible. Hence m unequal values of the generic expression Δ are obtained by changing C_1 successively into C_1, C_2, etc., namely,

$$d_1 C_1 + d_2 C_2 + \ldots + d_m C_m,$$
$$d_m C_1 + d_1 C_2 + \ldots + d_{m-1} C_m,$$
$$\ldots\ldots\ldots\ldots\ldots\ldots\ldots$$
$$d_2 C_1 + d_3 C_2 + \ldots + d_1 C_m.$$

To show that these expressions are all unequal, take the first two. If these were equal, we should have

$$(d_m - d_1) C_1 + (d_1 - d_2) C_2 + \text{etc.} = 0.$$

Therefore, by §13, $d_m - d_1 = 0$, $d_1 - d_2 = 0$, and so on; which, because d_1, d_2, etc., are not all equal to one another, is impossible. Since then Δ has at least m unequal particular cognate forms, Δ_1 is, by Prop. III., the root of a rational irreducible equation of a degree not lower than the mth; which, by Prop. VII., is impossible. Therefore k_1, k_2, etc., are rational. Hence each of the expressions in (27) is a rational function of t_1.

§37. *Cor.* Any expression of the type $k_1 + k_2 t_1 + k_3 t_1^2 +$ etc., which is such that all the unequal particular cognate forms of the generic expression under which it falls are obtained by substituting for t_1 successively the different primitive m^{th} roots of unity, while k_1, k_2, etc., remain unaltered, is a rational function of t_1. For, in the Proposition, Δ_1 or $k_1 + k_2 t_1 +$ etc. was shown to be a rational function of t_1, the conclusion being based on the circumstance that Δ_1 satisfies the condition specified.

§38. Proposition IX. If g be the sum of the roots of the equation $F(x) = 0$,

$$r_2 = \frac{1}{m}\left(g + \Delta_1^{\frac{1}{m}} + a_1 \Delta_1^{\frac{2}{m}} + b_1 \Delta_1^{\frac{3}{m}} + \ldots\right.$$
$$\left. + e_1 \Delta_1^{\frac{m-2}{m}} + h_1 \Delta_1^{\frac{m-1}{m}}\right); \qquad (29)$$

For, z being one of the whole numbers, 1, 2, ..., $m - 1$, put

$$p_z = (r_1 + t_1^z r_2 + t_1^{2z} r_3 + \text{etc.})(r_1 + t_1 r_2 + t_1^2 r_3 + \text{etc.})^{-z}. \qquad (30)$$

Multiply the first of its factors by t_1^{-z} and the second by t_1^z. Then

$$p_z = (r_2 + t_1^z r_3 + t_1^{2z} r_4 + \text{etc.})(r_2 + t_1 r_3 + t_1^2 r_4 + \text{etc.})^{-z}. \qquad (31)$$

Hence p_z does not alter its value when we change r_1 into r_2, r_2 into r_3, and so on. In like manner it does not alter its value when we change r_1 into r_a, r_2 into r_{a+1}, and so on. Therefore, by §33, p_z is not changed by any alterations that may be made in r_1, r_2, etc., while t_1 remains unaltered. Consequently, if p_z be a particular cognate form of P, all the unequal particular cognate forms of P are obtained by substituting for t_1 successively in p_z the different primitive m^{th} roots of unity, while r_1, r_2, etc., remain unaltered. Therefore, by Cor., Prop. VIII., p_z is a rational function of t_1. When $z = 2$, let $p_z = a_1$; when $z = 3$, let $p_z = b_1$, and so on. Then, from (26) and (30), $\Delta_2^{\frac{1}{m}} = a_1 \Delta_1^{\frac{2}{m}}$, $\Delta_3^{\frac{1}{m}} = b_1 \Delta_1^{\frac{3}{m}}$ and so on. But, from (27), since g is the sum of the roots of the equation $F(x) = 0$,

$$r_1 = \frac{1}{m}\left(g + \Delta_1^{\frac{1}{m}} + \Delta_2^{\frac{1}{m}} + \ldots\ldots + \Delta_{m-1}^{\frac{1}{m}}\right).$$

By putting $a_1 \Delta_1^{\frac{2}{m}}$ for $\Delta_2^{\frac{1}{m}}$, $b_1 \Delta_1^{\frac{3}{m}}$ for $\Delta_3^{\frac{1}{m}}$ and so on, this becomes (29). Because a_1, b_1, etc., are rational functions of t_1, while Δ_1, the root of a rational irreducible equation of the $(m-1)^{\text{th}}$ degree, is also a rational function of t_1, the coefficients a_1, b_1, etc., involve no surd that is not subordinate to $\Delta_1^{\frac{1}{m}}$.

§39. Proposition X. If the prime number m be odd, the expressions

$$\Delta_1^{\frac{1}{m}} \Delta_{m-1}^{\frac{1}{m}},\ \Delta_2^{\frac{1}{m}} \Delta_{m-2}^{\frac{1}{m}},\ \ldots\ldots,\ \Delta_{\frac{m-1}{2}}^{\frac{1}{m}} \Delta_{\frac{m+2}{m}}^{\frac{1}{m}}, \qquad (32)$$

are the roots of a rational equation of the $\left(\frac{m-1}{2}\right)^{\text{th}}$ degree.

By §32, when r_1, is charged into r_z, r_2 becomes r_{z+1}, r_3 becomes r_{z+2}, and so on. Hence the terms $r_1 r_2$, $r_2 r_3$, ... $r_m r_1$, form a cycle, the sum of the terms in which may be denoted by the symbol Σ_2^1. In like manner the sum of the terms in the cycle $r_1 r_3$, $r_2 r_4$, ..., $r_m r_2$, may be written Σ_3^1. And so on. In harmony with this notation, the sum of the m terms r_1^2, r_2^2, etc., may be written Σ_1^1. Now r_1 can only be changed into one of the terms r_1, r_2, etc.; and we have seen that, when it becomes r_z; r_2 becomes r_{z+1}, and so on. Such changes leave the cycle $r_1 r_2$, $r_2 r_3$, etc., as a whole unaltered.

Therefore, by Prop. III., Σ_2^1 is the root of a simple equation, or has a rational value. In like manner each of the expressions

$$\Sigma_1^1\,,\ \Sigma_2^1\,,\ \Sigma_3^1\,,\ \ldots\ldots\,,\ \Sigma_m^2\,, \tag{33}$$

has a rational value. From (26), by actual multiplication,

$$\Delta_1^{\frac{1}{m}}\,\Delta_{m-1}^{\frac{1}{m}} = \Sigma_1^1 + (\Sigma_2^1)\,t_1 + (\Sigma_3^1)\,t_1^2 + \text{etc.}$$

But Σ_2^1, Σ_3^1, etc., are respectively identical with Σ_m^1, Σ_{m-1}^1, etc. Therefore

$$\Delta_1^{\frac{1}{m}}\,\Delta_{m-1}^{\frac{1}{m}} = \Sigma_1^1 + (\Sigma_2^1)\,(t_1 + t_1^{-1}) + (\Sigma_3^1)(t_1^2 + t_1^{-2}) + \text{etc.} \tag{34}$$

Hence, since the terms in (33) are all rational, and since the terms in (32) are respectively what $\Delta_1^{\frac{1}{m}}\,\Delta_{m-1}^{\frac{1}{m}}$ becomes by changing t_1 successively into the $\frac{m-1}{2}$ terms t_1, t_1^2, etc., the terms in (32) are the roots of a rational equation of the $\left(\frac{m-1}{2}\right)^{\text{th}}$ degree.

§40. For the solution of the equation $x^n - 1 = 0$, n being a prime number such that m is a prime measure of $n - 1$, it is necessary to obtain the solution of the equation of the mth degree which has for one of its roots the sum of the $\frac{n-1}{m}$ terms in a cycle of primitive n^{th} roots of unity. This latter equation will be referred to as the *reducing Gaussian equation* of the mth degree to the equation

$$x^n - 1 = 0\,.$$

§41. Proposition XI. When the equation $F(x) = 0$ is the reducing Gaussian (see §40) of the mth degree to the equation $x^n - 1 = 0$, each of the $\frac{m-1}{2}$ expressions in (32) is equal to n.

Let the sum of the primitive n^{th} roots of unity forming the cycle (8), which sum has in preceding sections been indicated by the symbol C_1, be the root r_1 of the equation $F(x) = 0$. This implies, since s is the number of the terms in (8), that $ms = n - 1$. Let us reason first on the assumption that the cycle (8) is made up of pairs of reciprocal roots ω_1 and ω_1^{-1}, and so on. Then, because the cycle consists of $\frac{s}{2}$ pairs of reciprocal roots, C_1^2 or r_1^2 is the sum of

s^2 terms, each an nth root of unity. Among these unity occurs s times. Let ω_1 occur h_1 times; and let ω_1^λ the second term in (8), occur h' times. Since ω_1^λ may be made the first term in the cycle (8), it must, under the new arrangement, present itself in the value of r_1^2, precisely where ω_1 previously appeared. That is to say, $h' = h_1$. In like manner each of the terms in (8) occurs exactly h_1 times in the expression for r_1^2. The cycle (9) being that which contains all the primitive n^{th} roots of unity, let us, adhering to the notation of previous sections, suppose that, when ω_1 is changed into ω_1^β, C_1 or r_1 becomes C_2 or r_2, C_2 or r_2 becomes C_3 or r_3, and so on. On the same grounds on which every term in (8) occurs the same number of times in the value of r_1^2, each term in the cycle of terms whose sum is C_2 occurs the same number of times; and so on. Therefore

$$r_1^2 = s + h_1 C_1 + h_2 C_2 + \ldots + h_m C_m.$$

$$r_2^2 = s + h_m C_1 + h_1 C_2 + \ldots + h_{m-1} C_m,$$

$$\ldots\ldots\ldots\ldots\ldots\ldots\ldots\ldots\ldots\ldots\ldots\ldots$$

$$r_m^2 = s + h_2 C_1 + h_3 C_2 + \ldots + h_1 C_m.$$

Therefore, keeping in view (11), $\Sigma_1^1 = ms - (h_1 + h_2 + \ldots + h_m)$. But $s^2 - s$ is the number of the terms in the value of r_1^2 which are primitive n^{th} roots of unity. And this must be equal to

$$s\,(h_1 + \ldots + h_m).$$

Therefore

$$h_1 + h_2 + \ldots + h_m = s - 1 \;\therefore\; \Sigma_1^1 = ms + 1 - s = n - s.$$

Again, because r_1 is made up of pairs of reciprocal roots, and because therefore unity does not occur among the s^2 terms of which $r_1 r_2$ is the sum,

$$r_1 r_2 = k_1 C_1 + k_2 C_2 + \ldots + k_m C_m,$$

$$r_2 r_3 = k_m C_1 + k_1 C_2 + \ldots + k_{m-1} C_m,$$

$$\ldots\ldots\ldots\ldots\ldots\ldots\ldots\ldots\ldots\ldots\ldots\ldots$$

$$r_m r_1 = k_2 C_1 + k_3 C_2 + \ldots + k_1 C_m;$$

where k_1, k_2, etc., are whole numbers whose sum is s. Therefore $\Sigma_2^1 = -s$. In like manner each of the terms in (33) except the first is equal to $-s$. Therefore (34) becomes

$$\Delta_1^{\frac{1}{m}} \Delta_{m-1}^{\frac{1}{m}} = (n - s) - s\,(t_1 + t_1^2 + \text{etc.}) = n.$$

Let us reason now on the assumption that the cycle (8) is not made up of pairs of reciprocal roots. It contains in that case no reciprocal roots. By the same reasoning as above we get $\Sigma^1_1 = -s$. As regards the terms in (33) after the first, one of the terms C_1, C_2, etc., say C_z, must be such that the nth roots of unity of which it is the sum are reciprocals of those of which C_1 is the sum. In passing from C_1 to C_z, we change r_1 into r_z. In fact, C_1 being r_1, C_z is r_z. This being kept in view, we get, by the same reasoning as above, $\Sigma^1_z = n - s$. But, if any of the expressions C_1, C_2, etc., except C_z be selected, say C_a, none of the roots in (8) are reciprocals of any of those of which C_a is the sum. Therefore $\Sigma^1_a = -s$. Therefore, from (34)

$$\Delta_1^{\frac{1}{m}} \Delta_{m-1}^{\frac{1}{m}} = -s + (n-s)\, t_1^{z-1}$$

$$-s \left\{ (t_1 + t_1^2 + \ldots + t_1^{m-1}) - t_1^{z-1} \right\} = n.$$

In like manner every one of the expressions in (34) can be shown to have the value n.

§42. Two numerical illustrations of the law established in the preceding section may be given. The reducing Gaussian equation of the third degree to the equation $x^{19} - 1 = 0$ is $x^3 - x^2 - 6x - 7 = 0$; which gives

$$r_1 = \tfrac{1}{3}\left(-1 + \Delta_1^{\frac{1}{3}} + \Delta_2^{\frac{1}{3}}\right),$$

$$2\Delta_1 = 19\,(7 + 3\sqrt{3}),$$

$$2\Delta_2 = 19\,(7 - 3\sqrt{3}),$$

$$\Delta_1^{\frac{1}{3}} \Delta_2^{\frac{1}{3}} = 19.$$

The next example is taken from Lagrange's Theory of Algebraical Equations, Note XIV., §30. The Gaussian of the fifth degree to the equation $x^{11} - 1 = 0$ is $x^5 + x^4 - 4x^3 - 3x^2 + 3x + 1 = 0$; which gives

$$r_1 = \tfrac{1}{5}\left(-1 + \Delta_1^{\frac{1}{5}} + \Delta_2^{\frac{1}{5}} + \Delta_3^{\frac{1}{5}} + \Delta_4^{\frac{1}{5}}\right);$$

$$4\,\Delta_1 = 11\,(-89 - 25\sqrt{5} + 5p - 45q),$$

$$4\,\Delta_2 = 11\,(-89 + 25\sqrt{5} - 45p - 5q),$$

$$4\,\Delta_4 = 11\,(-89 - 25\sqrt{5} - 5p + 45q),$$

$$4\,\Delta_3 = 11\,(-89 + 25\sqrt{5} + 45p + 5q),$$

$$p = \sqrt{(-5 - 2\sqrt{5})},$$

$$q - \sqrt{(-5 + 2\sqrt{5})},$$

$$pq = -\sqrt{5} \;\therefore\; \Delta_1 \Delta_4 = 11^5.$$

§43. Proposition XII. To solve the Gaussian.

The path we have been following leads directly, assuming the primitive m^{th} root of unity t_1 to be known, to the solution of the reducing Gaussian equation of the mth degree to the equation $x^n - 1 = 0$. For, as in §41, the roots of the Gaussian are C_1, C_2, etc. Therefore g, the sum of the roots, is -1. Therefore

$$r_1 = \frac{1}{m}(-1 + \Delta_1^{\frac{1}{m}} + \Delta_2^{\frac{1}{m}} + \dots + \Delta_{m-1}^{\frac{1}{m}}). \qquad (35)$$

By Prop. VIII., Δ_1, Δ_2, etc., are rational functions of t_1. Therefore

$$\left.\begin{aligned}
\Delta_1 &= k_1 + k_2 t_1 + k_3 t_1^{2} + \dots + k_m t_1^{m-1} \\
\Delta_2 &= k_1 + k_2 t_1^{2} + k_3 t_1^{4} + \dots + k_m t_1^{2(m-1)} \\
&\dots\dots\dots\dots\dots\dots\dots\dots\dots \\
\Delta_{m-1} &= k_1 + k_2 t_1^{-1} + k_3 t_1^{-2} + \dots + k_m t_1 ;
\end{aligned}\right\} \qquad (36)$$

where k_1, k_2, etc., are rational. From the first of equations (26), putting C_1 for r_1, C_2 for r_2, and so on,

$$\Delta_1 = (C_1 + t_1 C_2 + \text{etc.})^m.$$

By actual involution this gives us k_1, k_2, etc., as determinate functions of C_1, C_2, etc., and therefore as known rational quantities. For instance take k_1. Being a determinate function of C_1, C_2, etc., we have

$$k_1 = q_1 + q_2 C_1 + q_3 C_2 + \dots + q_m C_{m-1};$$

where q_1, q_2, etc., are known rational quantities. But, by §13, the rational coefficients $q_1 - k_1$, q_2, etc., are all equal to one another. Therefore $k_1 = q_1 - q_2$. In like manner k_2, k_3, etc., are known. Therefore, from (36), Δ_1, Δ_2, etc., are known. Therefore, from (35), r_1 is known.

§44. Proposition XIII. The law established in Prop. X falls under the following more general law. The $m - 1$ expressions in each of the groups

$$\left.\begin{aligned}
&(\Delta_1^{\frac{1}{m}} \Delta_{m-1}^{\frac{1}{m}}, \ \Delta_2^{\frac{1}{m}} \Delta_{m-2}^{\frac{1}{m}}, \dots, \Delta_{m-1}^{\frac{1}{m}} \Delta_1^{\frac{1}{m}},) \\
&(\Delta_1^{\frac{2}{m}} \Delta_{m-2}^{\frac{1}{m}}, \ \Delta_2^{\frac{2}{m}} \Delta_{m-4}^{\frac{1}{m}}, \dots, \Delta_{m-1}^{\frac{2}{m}} \Delta_2^{\frac{1}{m}},) \\
&(\Delta_1^{\frac{3}{m}} \Delta_{m-3}^{\frac{1}{m}}, \ \Delta_2^{\frac{3}{m}} \Delta_{m-6}^{\frac{1}{m}}, \dots, \Delta_{m-1}^{\frac{3}{m}} \Delta_3^{\frac{1}{m}};)
\end{aligned}\right\} \qquad (37)$$

and so on, are the roots of a rational equation of the $(m - 1)^{\text{th}}$ degree.

The $m - 1$ terms in the first of the groups (37) are the $\frac{m-1}{2}$ terms in (32) each taken twice. Therefore, by Prop. X., the law enunciated in the present Proposition is established so far as this groupe is concerned. The general proof is as follows. By (30) in §38, taken in connection with (26), $p_{m-z}\, \Delta_1^{\frac{m-z}{m}} = \Delta_{m-z}^{\frac{1}{m}}$. Therefore $\Delta_1^{\frac{z}{m}}\, \Delta_{m-s}^{\frac{1}{m}} = p_{m-z}\, \Delta_1$. But, by §38, p_{m-s} is a rational function of t_1; and, by Prop. VIII., Δ_1 is a rational function of t_1. Therefore $\Delta_1^{\frac{z}{m}}\, \Delta_{m-s}^{\frac{1}{m}}$ is a rational function of t_1. Also from the manner in which p_{m-s} is formed, when t_1 in $p_{m-2}\, \Delta_1$ is changed sucessively into $t_1\, t_1^2, \ldots, t_1^{m-1}$, the expression $\Delta_1^{\frac{s}{m}}\, \Delta_{m-s}^{\frac{1}{m}}$ is changed successively into the $m - 1$ terms of that one of the groups (37) whose first term is $\Delta_1^{\frac{z}{m}}\, \Delta_{m-z}^{\frac{1}{m}}$. Therefore the terms in that group are the roots of a rational equation.

§45. *Cor.* The law established in the Proposition may be brought under a yet wider generalization. The expression

$$\Delta_1^{\frac{a}{m}}\, \Delta_2^{\frac{b}{m}}\, \Delta_3^{\frac{c}{m}} \ldots \Delta_{m-1}^{\frac{s}{m}} \qquad (38)$$

is the root of a rational equation of the $(m - 1)^{\text{th}}$ degree, if

$$a + 2b + 3c + \ldots + (m - 1)\, s = Wm,$$

W being a whole number. For, by (30) in connection with (26), $\Delta_2^{\frac{1}{m}} = p_2\, \Delta_1^{\frac{2}{m}}$, $\Delta_3^{\frac{1}{m}} = p_3\, \Delta_3^{\frac{1}{m}}$, and so on. Therefore (38) has the value

$$(p_2^b\, p_3^c \ldots)\, \Delta_1^{\frac{a+2b+3c+\ldots+(m-1)s}{m}}, \text{ or } (p_2^b\, p_3^c \ldots)\, \Delta_1^W.$$

This is a rational function of t_1, and therefore the root of a rational equation of the $(m - 1)^{\text{th}}$ degree.

THE EQUATION $F(x) = 0$ OF THE SECOND CLASS.

§46. We now suppose that the simplified root r_1 of the rational irreducible equation $F(x) = 0$ of the mth degree, m prime, involves, when modified according to §21, a principal surd not a root of unity. It must not be forgotten that, when we thus speak of roots of unity, we mean, according to §1, roots which have prime numbers for the denominators of their indices. In this case conclusions can be established similar to those reached in the case that has been considered. The root r_1 is still of the form (29). The equation $F(x) = 0$ has still an auxiliary of the $(m - 1)^{\text{th}}$ degree, whose roots are the mth powers of the expressions

$$\Delta_1^{\frac{1}{m}},\ a_1\,\Delta_1^{\frac{2}{m}},\ b_1\,\Delta_1^{\frac{3}{m}},\ \ldots\ldots,\ e_1\,\Delta_1^{\frac{m-2}{m}},\ h_1\,\Delta_1^{\frac{m-1}{m}}, \tag{39}$$

though the auxiliary here is not necessarily irreducible. Also, substituting the expressions in (39) for $\Delta_1^{\frac{1}{m}}$ $\Delta_2^{\frac{1}{m}}$, etc., in (37), the law of Proposition XIII. still holds, together with corollary in §45.

§47. By Cor. Prop. VI., the denominator of the index of a surd of the highest rank in r_1 is m. Let $\Delta_1^{\frac{1}{m}}$ be such a surd. By §21, the coefficients of the different powers of $\Delta_1^{\frac{1}{m}}$ in r_1 cannot be all zero. We may take the coefficient of the first power to be distinct from zero and to be $\frac{1}{m}$ for, if it were $\frac{k_1}{m}$, we might substitute $s^{\frac{1}{m}}$ for $k_1\Delta_1^{\frac{1}{m}}$, and so eliminate $\Delta_1^{\frac{1}{m}}$ from r_1, introducing in its room the new surd $s^{\frac{1}{m}}$ with $\frac{1}{m}$ for the coefficient of its first power. We may then put

$$r_1 = \frac{1}{m}\left(g + \Delta_1^{\frac{1}{m}} + a_1\,\Delta_1^{\frac{3}{m}} + \ldots\ldots + e_1\,\Delta_1^{\frac{m-2}{m}} + h_1\,\Delta_1^{\frac{m-1}{m}}\right); \tag{40}$$

where g, a_1, etc., are clear of $\Delta_1^{\frac{1}{m}}$. When $\Delta_1^{\frac{1}{m}}$ is changed successively into $\Delta_1^{\frac{1}{m}}$, $t_1^{-1}\,\Delta_1^{\frac{1}{m}}$, $t_1^{-2}\,\Delta_1^{\frac{1}{m}}$, etc., let

$$r_1,\ r_2,\ \ldots\ldots\ r_m, \tag{41}$$

be respectively what r_1 becomes, t_1 being a primitive m^{th} root of unity. By Prop. VI., the terms in (41) are the roots of the equation $F(x) = 0$. Taking r_n, any one of the particular cognate forms of R, let $\Delta_n^{\frac{1}{m}}$, a_n, etc., be respectively what $\Delta_1^{\frac{1}{m}}$, a_1, etc., become in passing from r_1 to r_n; and when $\Delta_n^{\frac{1}{m}}$ is changed successively into the different m^{th} roots of the determinate base Δ_n, let r_n become

$$r_n,\ r_n',\ r_n'',\ \ldots\ldots,\ r_n^{(m-1)}. \tag{42}$$

By Prop. II., the terms in (42) are roots of the equation $F(x) = 0$; and, by §23, they are all unequal. Therefore they are identical, in some order, with the terms in (41). Also, the sum of the terms in (41) is g. Therefore g is rational.

§48. Proposition XIV. In r_1, as expressed in (40), $\Delta_1^{\frac{1}{m}}$ is the only principal (see §2) surd.

Suppose, if possible, that there is in r_1 a principal surd $z_1^{\frac{1}{c}}$ distinct from $\Delta_1^{\frac{1}{m}}$. And first, let $z_1^{\frac{1}{c}}$ be not a root of unity. (It will be kept in view that when, in such a case, we speak of roots of unity, the denominators of their indices are understood, according §1, to be prime numbers.) When $z_1^{\frac{1}{c}}$ is changed into $z_2^{\frac{1}{c}}$, one of the other c^{th} roots of z_1, let r_1, a_1, etc., become respectively r_1', a_1', etc. Then

$$mr_1' = g + \Delta_1^{\frac{1}{m}} + a_1'\ \Delta_1^{\frac{2}{m}} + \text{etc} \tag{43}$$

By Prop. II., r_1' is equal to a term in (41), say to r_n. And, by §48, putting t_{n-1} for t_1^{1-n},

$$mr_n = g + t_{n-1}\,\Delta_1^{\frac{1}{m}} + t_{n-1}^2\,a_1\,\Delta_1^{\frac{2}{m}} + \text{etc.} \tag{44}$$

Therefore,

$$\Delta_1^{\frac{1}{m}}\,(1 - t_{n-1}) + \Delta_1^{\frac{2}{m}}\,(a'_1 - a_1\,t_{n-1}^2) + \text{etc.} = 0. \tag{45}$$

This equation involves no surds except those found in the simplified expression r_1, together with the primitive m^{th} root of unity. Therefore the expression on the left of (45) is in a simple state. Therefore, by §8, the coefficients of the different powers of $\Delta_1^{\frac{1}{m}}$ are separately zero. Therefore $t_{n-1} = 1$, $a_1' = a_1$, $b_1' = b_1$, and so on. But, as was shown in Prop. V., $z_1^{\frac{1}{c}}$ being a principal surd not a root of unity in the simplified expression a_1, a_1' cannot be equal to a_1 unless $z_1^{\frac{1}{c}}$ can be eliminated from a_1 without the introduction of any new surd. In like manner b_1' cannot be equal to b_1 unless $z_1^{\frac{1}{c}}$ can be eliminated from b_1. And so on. Therefore, because $a_1' = a_1$, and $b_1' = b_1$, and so on, $z_1^{\frac{1}{c}}$ admits of being eliminated from r_1 without the introduction of any new surd, which, by §21, is impossible. Next, let $z_1^{\frac{1}{c}}$ be a root (see §1) of unity, which may be otherwise written θ_1. Let the different primitive c^{th} roots of unity be θ_1, θ_2, etc.; and, when θ_1 is changed successively into θ_1, θ_2, etc., let r_1 become successively r_1, r_1', etc. Suppose it possible that the $c - 1$ terms r_1, r_1', etc., are all equal. Since $z_1^{\frac{1}{c}}$ is a principal surd in r_1, we may put $r_1 = h\theta_1^{c-1} + k\theta_1^{c-2} + \ldots + l$; where h, k, etc., are clear of θ_1. Therefore $(c - 1)\, r_1 = cl - (h + k + \text{etc.})$ Thus $z_1^{\frac{1}{c}}$ may be eliminated from r_1 without the introduction of any new surd; which by §21 is impossible. Since then the terms r_1, r_1', etc., are not all equal, let r_1 and r_1' be unequal. Then r_1' is equal to a term in (41) distinct from r_1, say to r_n. Expressing mr_1 and mr_n as in (43) and (44), we deduce (45); which, as above, is impossible.

§49. Proposition XV. Taking r_1, r_n, $\Delta_n^{\frac{1}{m}}$, etc., as in §47, an equation

$$t\,\Delta_n^{\frac{1}{m}} = p\,\Delta_1^{\frac{c}{m}} \tag{46}$$

can be formed; where t is an mth root of unity, and c is a whole number less than m but not zero, and p involves only surds subordinate (see §3) to $\Delta_1^{\frac{1}{m}}$ or $\Delta_n^{\frac{1}{m}}$

By §47, one of the terms in (42) is equal to r_1. For our argument it is immaterial which be selected. Let $r_n = r_1$. Therefore

$$(h_n \Delta_n^{\frac{m-1}{m}} + e_n \Delta_n^{\frac{m-2}{m}} + \ldots + \Delta_n^{\frac{1}{m}})$$
$$- (h_1 \Delta_1^{\frac{m-1}{m}} + e_1 \Delta_1^{\frac{m-2}{m}} + \ldots + \Delta_1^{\frac{1}{m}}) = 0. \quad (47)$$

The coefficients of the different powers of $\Delta_n^{\frac{1}{m}}$ here are not all zero, for the coefficient of the first power is unity. Therefore by §5, an equation $t\Delta_n^{\frac{1}{m}} = l_1$ subsists, t being an mth root of unity, and l_1 involving only surds exclusive of $\Delta_n^{\frac{1}{m}}$ that occur in (47). By Prop. XIV., $\Delta_1^{\frac{1}{m}}$ is a surd of a higher rank (see §3) than any surd in (47) except $\Delta_n^{\frac{1}{m}}$. Therefore we may put

$$l_1 = d + d_1 \Delta_1^{\frac{1}{m}} + d_2 \Delta_1^{\frac{2}{m}} + \ldots + d_{m-1} \Delta_1^{\frac{m-1}{m}};$$

where d, d_1, etc., involve only surds lower in rank than $\Delta_1^{\frac{1}{m}}$. Then

$$\Delta_n = l_1^m = (d + d_1 \Delta_1^{\frac{1}{m}} + \text{etc.})^m$$
$$= d' + d_1' \Delta_1^{\frac{1}{m}} + d_2' \Delta_1^{\frac{2}{m}} + \text{etc.};$$

where d', d_1', etc., involve only surds lower in rank than $\Delta_1^{\frac{1}{m}}$. By §8, since $\Delta_1^{\frac{1}{m}}$ is a surd in the simplified expressions r_1, the coefficients $d' - \Delta_n$, d_1', etc., in the equation

5

$$(d' - \varDelta_n) + d_1' \varDelta_1^{\frac{1}{m}} + d_2' \varDelta_1^{\frac{1}{m}} + \text{etc.} = 0 \qquad (48)$$

are separately zero. Therefore $(d + d_1 \varDelta_1^{\frac{1}{m}} + \text{etc.})^m = d'$. And, t_1 being a primitive m^{th} root of unity,

$$(d + d_1 t_1 \varDelta_1^{\frac{1}{m}} + \text{etc.})^m = d' + d' t_1 \varDelta_1^{\frac{1}{m}} + \text{etc.} = d'.$$

Therefore,

$$(d + d_1 t_1 \varDelta_1^{\frac{1}{m}} + \text{etc.}) = t_1^a (d + d_1 \varDelta_1^{\frac{1}{m}} + d_2 \varDelta_1^{\frac{2}{m}} + \text{etc.}),$$

t_1^a being one of the mth roots of unity. In the same way in which the coefficients of the different powers of $\varDelta_1^{\frac{1}{m}}$ in (48) are separately zero, each of the expressions $d\,(1 - t_1^a)$, $d_1\,(t_1 - t_1^a)$, etc., must be zero. But not more than one of the $m - 1$ factors, $t_1 - t_1^a$, $t_1^2 - t_1^a$, etc., can be zero. Therefore not more than one of the $m - 1$ terms d_1, d_2, etc., is distinct from zero. Suppose if possible that all these terms are zero. Then $t\,\varDelta_n^{\frac{1}{m}} = d$. Therefore the different powers of $\varDelta_n^{\frac{1}{m}}$ can be expressed in terms of the surds involved in d and of the mth root of unity. Substitute for $\varDelta_n^{\frac{1}{m}}$, $\varDelta_n^{\frac{2}{m}}$ etc., in (47), their values thus obtained. Then (47) becomes

$$Q - (h_1 \varDelta_1^{\frac{m-1}{m}} + \ldots . + \varDelta_1^{\frac{1}{m}}) = 0; \qquad (49)$$

where Q involves no surds, distinct from the primitive m^{th} root of unity, that are not lower in rank than $\varDelta_1^{\frac{1}{m}}$; which, because the coefficient of the first power of $\varDelta_1^{\frac{1}{m}}$ in (49) is not zero, is, by §8, impossible. Hence there must be one, while at the same there can be only one of the $m - 1$ terms, d_1, d_2, etc., distinct from zero. Let

d_c be the term that is not zero. Then $t_1^s - t_1^a = 0$. Therefore $1 - t_1^a$ is not zero. Therefore $d = 0$. Therefore, putting p for d_c,

$$t \Delta_n^{\frac{1}{m}} = p \Delta_1^{\frac{c}{m}}.$$

§50. *Cor.* By the proposition, values of the different powers of $\Delta_n^{\frac{1}{m}}$ can be obtained as follows;

$$t \Delta_n^{\frac{1}{m}} = p \Delta_1^{\frac{c}{m}}, t^2 \Delta_n^{\frac{2}{m}} = q \Delta_1^{\frac{s}{m}}, t^3 \Delta_n^{\frac{3}{m}} = k \Delta_1^{\frac{z}{m}}, \text{etc.}; \qquad (50)$$

where p, q, etc., involve only surds that occur in Δ_1 or Δ_n; and c, s, z, etc., are whole numbers in the series $1, 2, \ldots, m-1$. No two of the numbers c, s, etc., can be the same; for they are the products, with multiples of the prime number m left out, of the terms in the series $1, 2, \ldots, m-1$, by the whole number c which is less than m. Therefore the series c, s, z, etc., is the series $1, 2, \ldots, m-1$, in a certain order.

§51. PROPOSITION XVI. If r_n be one of the particular cognate forms of R, the expressions

$$t \Delta_n^{\frac{1}{m}}, t^2 a_n \Delta_n^{\frac{2}{m}}, \ldots, t^{m-2} e_n \Delta_n^{\frac{m-2}{m}}, t^{m-1} h_n \Delta_n^{\frac{m-1}{m}}, \qquad (51)$$

are severally equal, in some order, to those in (39), t being one of the mth roots of unity.

By §47, one of the terms in (42) is equal to r_1. For our argument it is immaterial which be chosen. Let $r_n = r_1$. By Cor. Prop. XV., the equations (50) subsist. Substitute in (47) the values of the different powers of $\Delta_n^{\frac{1}{m}}$ so obtained. Then

$$(t^{-1} p \Delta_1^{\frac{c}{m}} + t^{-2} q a_n \Delta_1^{\frac{s}{m}} + \text{etc.})$$
$$- (\Delta_1^{\frac{1}{m}} + a_1 \Delta_1^{\frac{2}{m}} + \text{etc.}) = 0. \qquad (52)$$

By Cor. Prop. XV., the series $\Delta_1^{\frac{c}{m}}$, $\Delta_1^{\frac{s}{m}}$, etc., is identical, in some order, with the series $\Delta_1^{\frac{1}{m}}$, $\Delta_1^{\frac{2}{m}}$, etc. Also, by §8, since $\Delta_1^{\frac{1}{m}}$ is a surd occurring in the simplified expression r_1, and since besides $\varDelta_1^{\frac{1}{m}}$ there are in (52) no surds, distinct from the primitive m^{th} root of unity, that are not lower in rank than $\varDelta_1^{\frac{1}{m}}$, if the equation (52) were arranged according to the powers of $\varDelta_1^{\frac{1}{m}}$ lower than the mth, the coefficients of the different powers of $\varDelta_1^{\frac{1}{m}}$ would be separately zero. Hence $\varDelta_1^{\frac{1}{m}}$ is equal to that one of the expressions,

$$t^{-1} p \, \varDelta_1^{\frac{c}{m}}, \; t^{-2} q a_n \, \varDelta_1^{\frac{s}{m}}, \text{ etc.} \tag{53}$$

in which $\varDelta_1^{\frac{1}{m}}$ is a factor. In like manner $a_1 \varDelta_1^{\frac{2}{m}}$ is equal to that one of the expressions (53) in which $\varDelta_1^{\frac{2}{m}}$ is a factor. And so on. There fore the terms $\varDelta_1^{\frac{1}{m}}$, $a_1 \varDelta_1^{\frac{2}{m}}$, etc., forming the series (39), are severally equal, in some order, to the terms in (53), which are those forming the series (51.)

§52. Proposition XVII. The equation $F(x) = 0$ has a rational *auxiliary* (Compare Prop. VII.) equation $\varphi(x) = 0$, whose roots are the mth powers of the terms in (39).

Let the unequal particular cognate forms of the generic expression $\varDelta$ under which the simplified expression $\varDelta_1$ falls be

$$\varDelta_1, \varDelta_2, \ldots, \varDelta_e. \tag{54}$$

By Prop. XVI., there is a value t of the mth root of unity for which the expressions

$$t \, \varDelta_2^{\frac{1}{m}}, \; t^2 a_2 \varDelta_2^{\frac{2}{m}}, \ldots, \; t^{m-2} e_2 \varDelta_2^{\frac{m-2}{m}}, \; t^{m-1} h_2 \varDelta_2^{\frac{m-1}{m}} \tag{55}$$

are severally equal, in some order, to those in (39). Therefore $\varDelta_2$ is equal to one of the terms

$$\varDelta_1, \; a_1^m \varDelta_1^2, \ldots, e_1^m \varDelta_1^{m-2}, \; h_1^m \varDelta_1^{m-1}. \tag{56}$$

In like manner each of the terms in (54) is equal to a term in (56). And, because the terms in (54) are unequal, they are severally equal to different terms in (56). By Prop. III., the terms in (54) are the roots of a rational irreducible equation, say $\psi_1(x) = 0$. Rejecting from the series (56) the roots of the equation $\psi_1(x) = 0$, certain of the remaining terms must in the same way be the roots of a rational irreducible equation $\psi_2(x) = 0$. And so on. Ultimately, if $\varphi(x)$ be the continued product of the expressions $\psi_1(x)$, $\psi_2(x)$, etc., the terms in (56) are the roots of the rational equation $\varphi(x) = 0$.

§53. The equations $\psi_1(x) = 0$, $\psi_2(x) = 0$, etc., formed by means of the expressions $\psi_1(x)$, $\psi_2(x)$, etc., may be said to be *sub-auxiliary* to the equation $F(x) = 0$. It will be observed that the sub-auxiliaries are all irreducible.

§54. PROPOSITION XVIII. In passing from r_1 to r_n, while Δ_1 becomes Δ_n, the expressions a_1, b_1, which, by Prop. XIV., involve only surds occurring in Δ_1, must severally receive determinate values, a_n, b_n, etc. In other words, a_1 being a particular cognate form of A, there cannot, for the same value of Δ_n, be two particular cognate forms of A, as a_n and a_N, unequal to one another. And so in the case of b_1, e_1, etc.

For, just as each of the terms in (42) is equal to a term in (41), there are primitive m^{th} roots of unity τ and T such that the expressions

$$\tau \Delta_n^{\frac{1}{m}} + \tau^2 a_n \Delta_n^{\frac{2}{m}} + \text{etc.}, \quad T \Delta_N^{\frac{1}{m}} + T^2 a_N \Delta_N^{\frac{1}{m}} + \text{etc.},$$

are equal to one another. Therefore, if $\Delta_N = \Delta_n$, in which case, by assigning suitable values to τ and T, $\Delta_N^{\frac{1}{m}}$ may be taken to be equal to $\Delta_n^{\frac{1}{m}}$,

$$\Delta_n^{\frac{1}{m}}(\tau - T) + \Delta_n^{\frac{2}{m}}(a_n \tau^2 - a_N T^2) + \text{etc.} = 0. \qquad (57)$$

Suppose if possible that the coefficients of the different powers of $\Delta_1^{\frac{1}{m}}$ in (57) are not all zero. Then, by §5, $t\,\Delta_n^{\frac{1}{m}} = l_1$; t being an mth root of unity; and l_1 involving only surds of lower ranks than $\Delta_1^{\frac{1}{m}}$. Hence, by Prop. XV. and Cor. Prop. XV, $\Delta_1^{\frac{1}{m}}$ is a rational function of surds of lower ranks than $\Delta_1^{\frac{1}{m}}$ and of the primitive m^{th} root of unity; which, by the definition in §6, is impossible. Since then the coefficients of the different powers of $\Delta_n^{\frac{1}{m}}$ in (57) are separately zero, $\tau = T$, $a_n \tau^2 = a_N T^2$, therefore $a_n = a_N$.

§55. Proposition XIX. Let the terms in (39) be written respectively

$$\Delta_1^{\frac{1}{m}}, \delta_2^{\frac{1}{m}}, \delta_3^{\frac{1}{m}}, \ldots, \delta_{m-1}^{\frac{1}{m}}. \qquad (58)$$

The symbols Δ_1, δ_2, δ_3, etc., are employed instead of Δ_1, Δ_2, Δ_3, etc., because this latter notation might suggest, what is not necessarily true, that the terms in (56) are all of them particular cognate forms of the generic expression under which Δ_1 falls. Then (compare Prop. XIII.) the $m - 1$ expressions in each of the groups

$$\left.\begin{array}{l} (\Delta_1^{\frac{1}{m}} \delta_{m-1}^{\frac{1}{m}}, \delta_2^{\frac{1}{m}} \delta_{m-2}^{\frac{1}{m}}, \delta_3^{\frac{1}{m}} \delta_{m-3}^{\frac{1}{m}}, \ldots, \delta_{m-1}^{\frac{1}{m}} \Delta_1^{\frac{1}{m}},) \\ (\Delta_1^{\frac{2}{m}} \delta_{m-2}^{\frac{1}{m}}, \delta_2^{\frac{2}{m}} \delta_{m-4}^{\frac{1}{m}}, \delta_3^{\frac{2}{m}} \delta_{m-6}^{\frac{1}{m}}, \ldots, \delta_{m-1}^{\frac{2}{m}} \delta_2^{\frac{1}{m}},) \\ (\Delta_1^{\frac{3}{m}} \delta_{m-3}^{\frac{1}{m}}, \delta_2^{\frac{3}{m}} \delta_{m-6}^{\frac{1}{m}}, \delta_3^{\frac{3}{m}} \delta_{m-9}^{\frac{1}{m}}; \ldots, \delta_{m-1}^{\frac{3}{m}} \delta_3^{\frac{1}{m}},) \end{array}\right\} \qquad (59)$$

and so on, are the roots of a rational equation of the $(m - 1)^{th}$ degree. Also (compare Prop. X.) the first $\frac{m-1}{2}$ terms in the first of the groups (59) are the roots of a rational equation of the $\left(\frac{m-1}{2}\right)^{th}$ degree.

In the enunciation of the proposition the remark is made that the series (54) is not necessarily identical with the series

$$\Delta_1, \delta_2, \delta_3, \ldots, \delta_{m-1}.$$

The former consists of the unequal particular cognate forms of Δ; the latter consists of the roots of the auxiliary equation $\varphi(x) = 0$. These two series are identical only when the auxiliary is irreducible. To prove the first part of the proposition, take the terms forming the second of the groups (59). Because $\delta_{m-2}^{\frac{1}{m}}$ represents $e_1 \Delta_1^{\frac{m-2}{m}}$,

$$e_1 \Delta_1 = \Delta_1^{\frac{2}{m}} \delta_{m-}^{\frac{1}{m}} .$$

Let E be the generic symbol under which the simplified expression e_1 falls. By Prop. XVIII., when Δ_1 is changed successively into the c terms in (54), e_1 receives successively the determinate values $e_1, e_2, \ldots, e_c$; and therefore $e_1 \Delta_1$ receives successively the determinate values

$$e_1 \Delta_1 , \; e_2 \Delta_2 , \; \ldots , \; e_c \Delta_c . \tag{60}$$

There is therefore no particular cognate form of $E\Delta$ that is not equal to a term in (60). By Prop. XVI. there is a value of the mth root of unity t for which the terms in (55) are severally equal, in some order, to those in (39). Let the term in (39) to which $t\, \Delta_2^{\frac{1}{m}}$ is equal be $q_1 \Delta_1^{\frac{n}{m}}$ Then, applying the principle of Cor. Prop. XV., as in Prop. XVI., it follows that the term in (39) to which $t^{m-2} e_2 \Delta_2^{\frac{m-2}{m}}$ in (55) is equal is $k_1 \Delta_1^{\frac{M-2n}{m}}$, M being a multiple of m, and $M - 2n$ being less than m. Therefore $e_2 \Delta_2$ is equal to $q_1^2 k_1 \Delta_1^{\frac{M}{m}}$, which is the product of two of the terms in (39) occuring respectively at equal distances from opposite extremities of the series. In other words, $e_2 \Delta_2$ is equal to an expression $\delta_m^{\frac{2}{m}} \delta_{m-2n}^{\frac{1}{m}}$ in the second of the groups (59). In like manner every term in (60) is equal to an expression in the second of the groups (59). Let the unequal terms in (60) be

$$e_1 \Delta_1 , \text{ etc.} \tag{61}$$

Then, by Prop. III., the terms in (61) are the roots of a rational irreducible equation, say $f_1(x) = 0$. Rejecting these, which are distinct terms in the second of the groups (59), it can in like manner be shown that certain other terms in that group are the roots of a rational irreducible equation, say $f_2(x) = 0$. And so on. Ultimately, if $f(x)$ be the continued product of the expressions $f_1(x)$, $f_2(x)$, etc., the terms forming the second of the groups (59) are the roots of a rational equation of the $(m-1)^{\text{th}}$ degree. The proof applies substantially to each of the other groups. To prove the second part, it is only necessary to observe that, in the first of the groups (59), the last term is identical with the first, the last but one with the second, and so on.

§56. *Cor.* 1. The reasoning in the proposition proceeds on the assumption that the prime number m is odd. Should m be even, the series Δ_1, δ_1, etc., is reduced to its first term. The law may be considered even then to hold in the following form. The product $\Delta_1^{\frac{1}{m}} \Delta_1^{\frac{1}{m}}$ is the root of a rational equation of the $(m - 1)^{th}$ degree, or is rational. For this product is Δ_1, which, by Prop. XVII., is the root of an equation of the $(m - 1)^{th}$ degree.

§56. *Cor.* 2. I merely notice, without farther proof, that the generalization in §45 in the case when the equation $F(x) = 0$ is of the first (see §30) class holds in the present case likewise.

Analysis of Solvable Equations of the Fifth Degree.

§58. Let the solvable irreducible equation of the mth degree, which we have been considering, be of the fifth degree. Then, by Prop. IX. and §47, whether the equation belongs to the first or to the second of the two classes that have been distinguished, assuming the sum of the roots g to be zero,

$$r_1 = \tfrac{1}{5}\,(\Delta_1^{\frac{1}{5}} + a_1\,\Delta_1^{\frac{2}{5}} + e_1\,\Delta_1^{\frac{3}{5}} + h_1\,\Delta_1^{\frac{4}{5}}), \qquad (62)$$

though, when the equation is of the first class, the root, as thus presented, is not in a simple state.

§59. Proposition XX. If the auxiliary biquadratic has a rational root Δ_1 not zero, all the roots of the auxiliary biquadratic are rational.

Because Δ_1 is rational, the auxiliary biquadratic $\varphi\,(x) = 0$ is not irreducible. Therefore, by Prop. VII., the equation $F\,(x) = 0$ is of the second (see §30) class. Therefore, by Prop. XIV., $\Delta_1^{\frac{1}{5}}$ is the only principal surd in r_1. Consequently, because Δ_1 is rational, a_1, e_1 and h_1 are rational. Therefore Δ_1, $a_1^5\,\Delta_1^2, e_1^5\,\Delta_1^3$, $h_1^5\,\Delta_1^4$, which are the roots of the auxiliary biquadratic, are rational.

§60. Proposition XXI. If the auxiliary biquadratic has a quadratic sub-auxiliary $\psi_1\,(x) = 0$ with the roots Δ_1 and Δ_2, then $\Delta_2 = h_1^5\,\Delta_1^4$, and $\Delta_1 = h_2^5\,\Delta_2^4$; and $h_1\,\Delta_1$ is rational.

As in §52, t being a certain fifth root of unity, each term in (55) is equal to a term in (39). The first term in (55) cannot be equal to the first in (39), for this would make $\Delta_2 = \Delta_1$. Suppose if possible that the first in (55) is equal to the second in (39). Then, by equations (50), applied as in Prop. XVI.,

$$\left.\begin{aligned} t\,\Delta_2^{\frac{1}{5}} &= a_1\,\Delta_1^{\frac{2}{5}}, & t^2 a_2\,\Delta_2^{\frac{2}{5}} &= h_1\,\Delta_1^{\frac{4}{5}}, \\ t^3 e_2\,\Delta_2^{\frac{3}{5}} &= \Delta_1^{\frac{1}{5}}, & t^4 h_2\,\Delta_2^{\frac{4}{5}} &= e_1\,\Delta_1^{\frac{3}{5}}, \\ \text{therefore } \Delta_2 &= a_1^5\,\Delta_1^2, & a_2^5\,\Delta_2^2 &= h_1^5\,\Delta_1^4, \\ e_2^5\,\Delta_2^3 &= \Delta_1, & h_2^5\,\Delta_2^4 &= e_1^5\,\Delta_1^3. \end{aligned}\right\} \qquad (63)$$

Now $a_1^5\,\Delta_1^2$, being equal to Δ_2, is a root of the equation $\psi'_1(x) = 0$. And $a_1^5\,\Delta_1^2$, involving only surds that occur in r_1, is in a simple state. Therefore, by Prop. III., $a_2^5\,\Delta_2^2$ is a root of the equation $\psi_1(x) = 0$. Therefore $h_1^5\,\Delta_1^4$, and therefore also $h_2^5\,\Delta_2^4$ or $e_1^5\,\Delta_1^3$, are roots of that equation. Hence all the terms

$$\Delta_1\,,\ a_1^5\,\Delta_1^2\,,\ e_1^5\,\Delta_1^3\,,\ h_1^5\,\Delta_1^4\,, \qquad (64)$$

are roots of the equation $\psi'_1(x) = 0$. But a_1, e_1, h_1, are all distinct from zero; for, by (63), if one of them was zero, all would be zero, and therefore $\Delta_1^{\frac{1}{5}}$ would be zero; which by §6, is impossible. From this it follows that no two terms in (64) are equal to one another; for taking $a_1^5\,\Delta_1^2$ and $e_1^5\,\Delta_1^3$, if these were equal, we should have $e_1\,t'\,\Delta_1^{\frac{1}{5}} = a_1$, t' being a fifth root of unity; which; which by §8, is impossible. This gives the equation $\psi'_1(x) = 0$ four unequal roots; which, because it is of the second degree, is impossible. Therefore the first term in (55) is not equal to the second in (39). In the same way it can be shown that it is not equal to the third. Therefore it must be equal to the fourth. In like manner the first in (39) is equal to the fourth in (55). Because then $t\,\Delta_2^{\frac{1}{5}} = h_1\,\Delta_1^{\frac{4}{5}}$, and $\Delta_1^{\frac{1}{5}} = t^4\,h_2\,\Delta_2^{\frac{4}{5}}$, $h_2\,\Delta_2 = h_1\,\Delta_1$. But, just as it was proved in §56 that, the roots of the sub-auxiliary $\psi'_1(x) = 0$ being the c terms Δ_1, Δ_2, etc., there is no particular cognate form of $E\Delta$ that is not a term in the series $e_1\,\Delta_1$, $e_2\,\Delta_2$, ..., $e_c\,\Delta_c$, it follows that, if h_1 be a particular cognate form of H, there is no particular cognate form of $H\Delta$ that is not equal to one of the terms $h_1\,\Delta_1$ and $h_2\,\Delta_2$. Hence, since $h_1\,\Delta_1 = h_2\,\Delta_2$, $H\Delta$ has no particular cognate form different in value from $h_1\,\Delta_1$. Therefore, by Prop. III., $h_1\,\Delta_1$ is rational.

§61. PROPOSITION XXII. The auxiliary biquadratic $\varphi(x) = 0$ either has all its roots rational, or has a sub-auxiliary (see §53) of the second degree, or is irreducible.

It will be kept in view that the sub-auxiliaries are, by the manner of their formation, irreducible. First, let the series (54), containing the roots of the sub-auxiliary $\psi_1(x) = 0$ consist of a single term Δ_1. Then, by Prop. III., Δ_1 is rational. Therefore, by Prop. XX., all the roots of the auxiliary are rational. Next, let the series (54) consist of the two terms Δ_1 and Δ_2 · By this very hypothesis, the auxiliary biquadratic has a quadratic sub-auxiliary. Lastly, let the series (54) contain more than two terms. Then it has the three terms Δ_1, Δ_2, Δ_3. We have shown that these must be severally equal to terms in (64). Neither Δ_2 nor Δ_3 is equal to Δ_1. They cannot both be equal to $h_1^{5} \Delta_1^{4}$. Therefore one of them is equal to one of the terms $a_1^{5} \Delta_1^{2}$, $e_1^{5} \Delta_1^{3}$. But in §60 it appeared that, if Δ_2 be equal either to $a_1^{5} \Delta_1^{2}$ or to $e_1^{5} \Delta_1^{3}$, all the terms in (64) are roots of the irreducible equation of which Δ_1 is a root. The same thing holds regarding Δ_3. Therefore, when the series (54) contains more than two terms, the irreducible equation which has Δ_1 for one of its roots has the four unequal terms in (64) for roots; that is to say, the auxiliary biquadratic is irreducible.

§62. Let $5u_1 = \Delta_1^{\frac{1}{5}}$, $5u_2 = a_1 \Delta_1^{\frac{2}{5}}$, $5u_3 = e_1 \Delta_1^{\frac{3}{5}}$, $5u_4 = h_1 \Delta_1^{\frac{4}{5}}$; and, n being any whole number, let S_n denote the sum of the nth powers of the roots of the equation $F(x) = 0$. Then

$$S_1 = 0; \; S_2 = 10(u_1 u_4 + u_2 u_3); \; S_3 = 15 \left\{ \Sigma(u_1 u_2^2) \right\};$$

$$S_4 = 20 \left\{ \Sigma(u_1^3 u_2) \right\} + 30(u_1^2 u_4^2 + u_2^2 u_3^2) + 120\, u_1 u_2 u_3 u_4;$$

$$S_5 = 5 \left\{ \Sigma(u_1^5) \right\} + 100 \left\{ \Sigma(u_1^3 u_3 u_4) \right\} + 150 \left\{ \Sigma(u_1 u_3^2 u_4^2) \right.;$$

where such an expression as $\Sigma(u_1 u_2^2)$ means the sum of all such terms as $u_1 u_2^2$; it being understood that, as any one term in the circle u_1, u_2, u_4, u_3, passes into the next, that next passes into its next, u_3 passing into u_1.

THE ROOTS OF THE AUXILIARY BIQUADRATIC ALL RATIONAL.

§63. Any rational values that may be assigned to Δ_1, a_1, e_1, and h_1 in r_1, taken as in (62), make r_1 the root of a rational equation of the fifth degree, for they render the values of S_1, S_2, etc., in §62, rational. In fact, $S_1 = 0$, $25 S_2 = 10 \Delta_1 (h_1 + a_1 e_1)$, and so on.

THE AUXILIARY BIQUADRATIC WITH A QUADRATIC SUB-AUXILIARY.

§64. PROPOSITION XXIII. In order that r_1, taken as in (62), may be the root of an irreducible equation $F(x) = 0$ of the fifth degree, whose auxiliary biquadratic has a quadratic sub-auxiliary, it must be of the form

$$r_1 = \tfrac{1}{5}\left\{(\Delta_1^{\frac{1}{5}} + \Delta_2^{\frac{1}{5}}) + (a_1 \Delta_1^{\frac{2}{5}} + a_2 \Delta_2^{\frac{2}{5}})\right\}; \qquad (65)$$

where Δ_1 and Δ_2 are the roots of the irreducible equation $\psi_1(x) = x^2 - 2px + q^5 = 0$; and $a_1 = b + d\sqrt{(p^2 - q^5)}$, $a_2 = b - d\sqrt{(p^2 - q^5)}$; p, b and d being rational; and the roots $\Delta_1^{\frac{1}{5}}$ and $\Delta_2^{\frac{1}{5}}$ being so related that $\Delta_1^{\frac{1}{5}} \Delta_2^{\frac{1}{5}} = q$.

By Prop. VII., when a quintic equation is of the first (see §30) class, the auxiliary biquadratic is irreducible. Hence, in the case we are considering, the quintic is of the second class. The quadratic sub-auxiliary may be assumed to be $\psi_1(x) = x^2 - 2px + k = 0$, p and k being rational. By Prop. XXI., the roots of the equation $\psi_1(x) = 0$ are Δ_1 and $h_1^5 \Delta_1^4$. Therefore $k = (h_1 \Delta_1)^5$; or, putting q for $h_1 \Delta_1$, $k = q^5$. By the same proposition, $h_1 \Delta_1$ is rational. Therefore q is rational. Hence $\psi_1(x)$ has the form specified in the enunciation of the proposition. Next, by Proposition XVI., there is a fifth root of unity t such that $t\,\Delta_2^{\frac{1}{5}} = h_1 \Delta_1^{\frac{4}{5}}$. If we take t to be unity, which we may do by a suitable interpretation of the symbol $\Delta_2^{\frac{1}{5}}$, $\Delta_2^{\frac{1}{5}} = h_1 \Delta_1^{\frac{4}{5}}$. This implies that $e_1 \Delta_1^{\frac{3}{5}} = a_2 \Delta_2^{\frac{2}{5}}$, a_2 being what a_1 becomes in passing from Δ_1 to Δ_2. Substituting these values of $e_1 \Delta_1^{\frac{3}{5}}$ and $h_1 \Delta_1^{\frac{4}{5}}$ in (62), we obtain the form of r_1 in (65), while at the same time $\Delta_1^{\frac{1}{5}} \Delta_2^{\frac{1}{5}} = h_1 \Delta_1 = q$. The forms of a_1 and a_2 have to be more accurately determined. By Prop. XIV., $\Delta_1^{\frac{1}{5}}$ is the only principal surd that r_1, as presented in (62), contains. Therefore a_1 involves no surd that does not occur in Δ_1; that is to say, $\sqrt{(p^2 - q^5)}$ is the only surd in a_1. Hence we may put $a_1 = b + d\sqrt{(p^2 - q^5)}$; b and d being rational. But a_2 is what a_1 becomes in passing from Δ_1 to Δ_2. And Δ_2 differs from Δ_1 only in the sign of the root $\sqrt{(p^2 - q^5)}$. Therefore

$$a_2 = b - d\sqrt{(p^2 - q^5)}.$$

§65. Any rational values that may be assigned to b, d, p and q in r_1, taken as in (65), make r_1 the root of a rational equation of the fifth degree; for they render the values of S_1, S_2, etc., in §62, rational. In fact, $S_1 = 0$, $25\,S_2 = 10\{q + q^2 b^2 - q^2 d^2 (p^2 - q^5)\}$, and so on.

The Auxiliary Biqadratic Irreducible.

§66. When the auxiliary biquadratic is irreducible, the unequal particular cognate forms of Δ are, by Prop. III., four in number, Δ_1, Δ_2, Δ_3, Δ_4. As explained in §55, because the equation $\varphi(x) = 0$ is irreducible, these terms are severaly identical with Δ_1, δ_2, δ_3, δ_4. Hence, putting $m = 5$, the first two terms in the first of the groups (59) may be written in the notation of (37),

$$\Delta_1^{\frac{1}{5}}\,\Delta_4^{\frac{1}{5}},\ \Delta_2^{\frac{1}{5}}\,\Delta_3^{\frac{1}{5}}\ ; \tag{66}$$

and the second and third groups may be written

$$\left.\begin{array}{l} (\Delta_1^{\frac{2}{5}}\,\Delta_3^{\frac{1}{5}},\ \Delta_2^{\frac{2}{5}}\,\Delta_1^{\frac{1}{5}},\ \Delta_3^{\frac{2}{5}}\,\Delta_4^{\frac{1}{5}},\ \Delta_4^{\frac{2}{5}}\,\Delta_2^{\frac{1}{5}}) \\ (\Delta_1^{\frac{3}{5}}\,\Delta_2^{\frac{1}{5}},\ \Delta_2^{\frac{3}{5}}\,\Delta_4^{\frac{1}{5}},\ \Delta_3^{\frac{3}{5}}\,\Delta_1^{\frac{1}{5}},\ \Delta_4^{\frac{3}{5}}\,\Delta_3^{\frac{1}{5}}). \end{array}\right\} \tag{67}$$

§67. Proposition XXIV. The roots of the auxiliary biquadratic equation $\varphi(x) = 0$ are of the forms

$$\left.\begin{array}{l} \Delta_1 = m + n\sqrt{z} + \sqrt{s},\ \Delta_2 = m - n\sqrt{z} + \sqrt{s_1}, \\ \Delta_4 = m + n\sqrt{z} - \sqrt{s},\ \Delta_3 = m - n\sqrt{z} - \sqrt{s_1}\ ; \end{array}\right\} \tag{68}$$

where $s = p + q\sqrt{z}$, and $s_1 = p - q\sqrt{z}$; m, n, z, p and q being rational; and the surd $\sqrt{s}$ being irreducible.

By Propositions XIII. and XIX., the terms in (66) are the roots of a quadratic. Therefore $\Delta_1\,\Delta_4$ and $\Delta_2\,\Delta_3$ are the roots of a quadratic. Suppose if possible that $\Delta_1\,\Delta_3$ is the root of a quadratic. By Propositions IX. and XIX., $\Delta_3^{\frac{1}{5}} = e_1\,\Delta_1^{\frac{3}{5}}$. Therefore $e_1^5\,\Delta_1^4$ is the root of a quadratic. From this it follows (Prop. III.) that there are not more than two unequal terms in the series,

$$e_1^5\,\Delta_1^4,\ e_2^5\,\Delta_2^4,\ e_3^5\,\Delta_3^4,\ e_4^5\,\Delta_4^4. \tag{69}$$

But suppose if possible that $e_1^5\,\Delta_1^4 = e_2^5\,\Delta_2^4$. Then, t being one of the fifth roots of unity, $te_1\,\Delta_1^{\frac{4}{5}} = e_2\,\Delta_2^{\frac{4}{5}}$ But, by Propositions IX. and XIX., $\Delta_2^{\frac{1}{5}} = h_1\,\Delta_1^{\frac{4}{5}}$. Therefore, $te_1\,\Delta_1^{\frac{4}{5}} = e_2\,h_1^4\,\Delta_1^3\,\Delta_1^{\frac{1}{5}}$. Therefore, by §8, $e_1 = 0$. Therefore one of the roots of the auxiliary biquadratic is zero; which because the auxiliary biquadratic is assumed to be irreducible, is impossible. Therefore $e_1^5 \Delta_1^4$ and $e_2^5 \Delta_2^4$ are unequal. In the same way all the terms in (69) can be shown to be unequal; which, because it has been proved that there are not more than two unequal terms in (69), is impossible. Therefore $\Delta_1 \Delta_3$ is not the root of a quadratic equation. Therefore the product of two of the roots, Δ_1 and Δ_4, of the auxiliary biquadratic is the root of a quadratic equation, while the product of a different pair, Δ_1 and Δ_3, is not the root of a quadratic. But the only forms which the roots of an irreducible biquadratic can assume consistently with these conditions are those given in (68).

§68. Proposition XXV. The surd $\surd s_1$ can have its value expressed in terms of $\surd s$ and $\surd z$.

By Propositions XIII. and XIX, the terms of the first of the groups (67) are the roots of a biquadratic equation. Therefore their fifth powers

$$\Delta_1^2 \Delta_3, \quad \Delta_2^2 \Delta_1, \quad \Delta_3^2 \Delta_4, \quad \Delta_4^2 \Delta_2, \tag{70}$$

are the roots of a biquadratic. From the values of Δ_1, Δ_2, Δ_3 and Δ_4 in (68), the values of the terms in (70) may be expressed as follows:

$$\left.\begin{aligned}
\Delta_1^2 \Delta_3 &= F + F_1 \surd z + (F_2 + F_3 \surd z) \surd s \\
&\quad + (F_4 + F_5 \surd z) \surd s_1 + (F_6 + F_7 \surd z) \surd s \surd s_1, \\
\Delta_2^2 \Delta_1 &= F - F_1 \surd z + (F_2 - F_3 \surd z) \surd s_1 \\
&\quad - (F_4 - F_5 \surd z) \surd s - (F_6 - F_7 \surd z) \surd s \surd s_1, \\
\Delta_4^2 \Delta_2 &= F - F_1 \surd z - (F_2 - F_3 \surd z) \surd s_1 \\
&\quad + (F_4 - F_5 \surd z) \surd s - (F_6 - F_7 \surd z) \surd s \surd s_1, \\
\Delta_3^2 \Delta_4 &= F + F_1 \surd z - (F_2 + F_3 \surd z) \surd s \\
&\quad -(F_4 + F_5 \surd z) \surd s_1 + (F_6 + F_7 \surd z) \surd s \surd s_1,
\end{aligned}\right\} \tag{71}$$

where F, F_1, etc., are rational. Let $\Sigma (\Delta_1^2 \Delta_3)$ be the sum of the four expressions in (70). Then, because these expressions are the roots of a biquadratic, $\Sigma (\Delta_1^2 \Delta_3)$ or $4F + 4F_7 \surd s \surd s_1$, must be rational. Suppose if possible that $\surd s_1$ cannot have its value expressed in terms of $\surd s$ and $\surd z$. Then, because $\surd s \surd s_1$ is not rational, $F_7 = 0$. By (68), this implies that $n = 0$. Let

$$\begin{aligned}
(\Delta_1^2 \Delta_3)^2 &= L + L_1 \surd z + (L_2 + L_3 \surd z) \surd s \\
&\quad + (L_4 + L_5 \surd z) \surd s_1 + (L_6 + L_7 \surd z) \surd s \surd s_1,
\end{aligned}$$

where L, L_1, etc., are rational. Then, as above, $L_7 = 0$. Keeping in view that $n = 0$, this means that $m^2 q = 0$. But q is not zero, for this would make $\surd s = \surd s_1$; which, because we are reasoning on the hypothesis that $\surd s_1$ cannot have its value expressed in terms of $\surd s$ and $\surd z$, is impossible. Therefore m is zero. And it was shown that n is zero. Therefore $\Delta_1 = \surd s$, and $\Delta_3 = - \surd s$. Therefore $\Delta_1 \Delta_3 = - \surd (p^2 - q^2 z)$; which, because it has been proved that $\Delta_1 \Delta_3$ is not the root of a quadratic equation, is impossible. Hence $\surd s_1$ cannot but be a rational function of $\surd s$ and $\surd z$.

§69. Proposition XXVI. The form of s is

$$h (1 + e^2) + h \surd(1 + e^2), \tag{72}$$

h and e being rational, and $1 + e^2$ being the value of z.

By Prop. XXV., $\surd s_1 = v + c \surd s$, v and c being rational functions of $\surd z$. Therefore $s_1 = v^2 + c^2 s + 2vc \surd s$. By Prop. XXIV., $\surd s$ is irreducible. Therefore $vc = 0$. But c is not zero, for this would make $\surd s_1 = v$, and thus $\surd s_1$ would be the root of a quadratic equation. Therefore $v = 0$, and $\surd s_1 = c \surd s = (c_1 + c_2 \surd z) \surd s$, c_1 and c_2 being rational. Therefore

$$\begin{aligned}\surd (ss_1) = \surd (p^2 - q^2 z) &= (c_1 + c_2 \surd z)(p + q \surd z) \\ &= (c_1 p + c_2 q z) + \surd z (c_1 q + c_2 p) = P + Q \surd z.\end{aligned}$$

Here, since $p^2 - q^2 z$ is rational, either $P = 0$ or $Q = 0$. As the latter of these alternatives would make $\surd (p^2 - q^2 z)$ rational, and therefore would make $\surd (p + q \surd z)$ or $\surd s$ reducible, it is inadmissible. Therefore $c_1 p + c_2 qz = 0$, and

$$\surd (p^2 - q^2 z) = (c_1 q + c_2 p) \surd z.$$

Now qz is not not zero, for this would make $\surd (ss_1) = \pm p$; which, because $\surd s$ is irreducible, is impossible. Therefore $c_2 = 0$. But, by hypothesis, $c_1 = 0$; therefore $\surd s_1$, which is equal to $(c_1 + c_2 \surd z) \surd s$, is zero; which is impossible. Hence c_1 cannot be zero. We may therefore put $ce = 1$, and $h (1 + e^2) = p$. Then $s = p + q \surd z = h (1 + e^2) + h \surd (1 + e^2)$. Having obtained this form, we may consider z to be identical with $1 + e^2$, q with h, and p with $h (1 + e^2)$.

§70. The reasoning in the preceding section holds good whether the equation $F(x) = 0$ be of the first (see §30) or of the second class. If we had had to deal simply with equations of the first class, the proof given would have been unnecessary, so far as the form of z is concerned; because, in that case, by Prop. VIII., Δ_1 is a rational function of the primitive fifth root of unity.

§71. PROPOSITION XXVII. Under the conditions that have been established, the root r_1 takes the form given without deduction in *Crelle* (Vol. V., p. 336) from the papers of Abel.

For, by *Cor.* Prop. XIII. (compare also *Cor.* 2, Prop. XIX.,) the expressions

$$\Delta_1^{\frac{1}{5}}\ \Delta_3^{\frac{2}{5}}\ \Delta_4^{\frac{4}{5}}\ \Delta_2^{\frac{3}{5}},\ \Delta_2^{\frac{1}{5}}\ \Delta_1^{\frac{2}{5}}\ \Delta_3^{\frac{4}{5}}\ \Delta_4^{\frac{3}{5}},$$

$$\Delta_3^{\frac{1}{5}}\ \Delta_4^{\frac{2}{5}}\ \Delta_2^{\frac{4}{5}}\ \Delta_1^{\frac{3}{5}},\ \Delta_4^{\frac{1}{5}}\ \Delta_2^{\frac{2}{5}}\ \Delta_1^{\frac{4}{5}}\ \Delta_3^{\frac{3}{5}}, \qquad (73)$$

are the roots of a biquadratic equation. In the corollaries referred to, it is merely stated that each of the expressions in (73) is the root of a biquadratic; but the principles of the propositions to which the corollaries are attached show that the four expressions must be the roots of the same biquadratic. Let the terms in (73) be denoted respectively by

$$5A_1^{-1}, \qquad 5A_2^{-1}, \qquad 5A_3^{-1}, \qquad 5A_4^{-1}.$$

Then $\Delta_1^{\frac{1}{5}}\ \Delta_3^{\frac{2}{5}}\ \Delta_4^{\frac{4}{5}}\ \Delta_2^{\frac{3}{5}} = \Delta_4^{\frac{5}{5}}\ (\Delta_1^{\frac{1}{5}}\ \Delta_3^{\frac{2}{5}}\ \Delta_4^{\frac{3}{5}}\ \Delta_2^{\frac{3}{5}})$ is an identity. Therefore

$$\tfrac{1}{5}\ \Delta_4^{\frac{1}{5}} = A_1\ (\Delta_1^{\frac{1}{5}}\ \Delta_3^{\frac{2}{5}}\ \Delta_4^{\frac{4}{5}}\ \Delta_2^{\frac{3}{5}}). \quad \text{Similarly,}$$

$$\tfrac{1}{5}\ \Delta_3^{\frac{1}{5}} = A_3\ (\Delta_3^{\frac{1}{5}}\ \Delta_4^{\frac{2}{5}}\ \Delta_2^{\frac{4}{5}}\ \Delta_1^{\frac{3}{5}})$$

$$\tfrac{1}{5}\ \Delta_2^{\frac{1}{5}} = A_2\ (\Delta_2^{\frac{1}{5}}\ \Delta_1^{\frac{2}{5}}\ \Delta_3^{\frac{4}{5}}\ \Delta_4^{\frac{3}{5}}), \text{ and}$$

$$\tfrac{1}{5}\ \Delta_1^{\frac{1}{5}} = A_4\ (\Delta_4^{\frac{1}{5}}\ \Delta_2^{\frac{2}{5}}\ \Delta_1^{\frac{4}{5}}\ \Delta_3^{\frac{3}{5}}).$$

Substituting these values in (62), we get

$$r_1 = A_1\ (\Delta_1^{\frac{1}{5}}\ \Delta_3^{\frac{2}{5}}\ \Delta_4^{\frac{4}{5}}\ \Delta_2^{\frac{3}{5}}) + A_2\ (\Delta_2^{\frac{1}{5}}\ \Delta_1^{\frac{2}{5}}\ \Delta_3^{\frac{4}{5}}\ \Delta_4^{\frac{3}{5}})$$

$$+ A_3\ (\Delta_3^{\frac{1}{5}}\ \Delta_4^{\frac{2}{5}}\ \Delta_2^{\frac{4}{5}}\ \Delta_1^{\frac{3}{5}}) + A_4\ (\Delta_4^{\frac{1}{5}}\ \Delta_2^{\frac{2}{5}}\ \Delta_1^{\frac{4}{5}}\ \Delta_3^{\frac{3}{5}}). \qquad (74)$$

This, with immaterial differences in the subscripts, is Abel's expression; only we need to determine A_1, A_2, A_3 and A_4 more exactly. These terms are the reciprocals of the terms in (73) severally divided by 5. Therefore they are the roots of a biquadratic. Also, no surds can appear in A_1 except those that are present in Δ_1, Δ_2, Δ_3 and Δ_4. That is to say, A_1 is a rational function of $\surd s$, $\surd s_1$ and $\surd z$. But it was shown that $\surd s_1 \surd s = he \surd z$. Therefore A_1 is a rational function of $\surd s$ and $\surd z$. We may therefore put

$$A_1 = K + K'\ \Delta_1 + K''\ \Delta_4 + K'''\ \Delta_1\ \Delta_4,$$

K, K', K'' and K''' being rational. But the terms A_1, A_2, A_4, A_3 circulate with Δ_1, Δ_2, Δ_4, Δ_3. Therefore

$$A_2 = K + K'\Delta_2 + K''\Delta_3 + K'''\Delta_2\Delta_3,$$
$$A_4 = K + K'\Delta_4 + K''\Delta_1 + K'''\Delta_1\Delta_4,$$
$$A_3 = K + K'\Delta_3 + K''\Delta_2 + K'''\Delta_2\Delta_3,$$

These are Abel's values.

§72. Keeping in view the values of Δ_1, Δ_2, etc., in (67), and also that $z = 1 + e^2$, and $s = hz + h\sqrt{z}$, any rational values that may be assigned to m, n, e, h, K, K', K'' and K''' make r_1, as presented in (74), the root of an equation of the fifth degree. For, any rational values of m, n, etc., make the values of S_1, S_2, etc., in §62, rational.

§73. It may be noted that, not only is the expression for r_1 in (74) the root of a quintic equation whose auxiliary biquadratic is irreducible, but on the understanding that the surds $\sqrt{s}$ and $\sqrt{z}$ in Δ_1 may be reducible, the expression for r_1 in (74) contains the roots both of all equations of the fifth degree whose auxiliary biquadratics have their roots rational, and of all that have quadratic sub-auxiliaries. It is unecessary to offer proof of this.

§74. The equation $x^5 - 10x^3 + 5x^2 + 10x + 1 = 0$ is an example of a solvable quintic with its auxiliary biquadratic irreducible. One of its roots is

$$\omega^{\frac{1}{5}} + \omega\omega^{\frac{2}{5}} + \omega^3\omega^{\frac{3}{5}} + \omega^4\omega^{\frac{4}{5}},$$

ω being a primitive fifth root of unity. It is obvious that this root satisfies all the conditions that have been pointed out in the preceding analysis as necessary. A root of an equation of the seventh degree of the same character is

$$\omega^{\frac{1}{7}} + \omega^4\omega^{\frac{2}{7}} + \omega^4\omega^{\frac{3}{7}} + \omega^2\omega^{\frac{4}{7}} + \omega^3\omega^{\frac{5}{7}} + \omega^6\omega^{\frac{6}{7}},$$

ω being a primitive seventh root of unity. The general form under which these instances fall can readily be found. Take the cycle that contains all the primitive $(m^2)^{\text{th}}$ roots of unity,

$$\theta, \theta^{\beta}, \theta^{\beta^2}, \text{ etc.} \qquad (75)$$

m being prime. The number of terms in the cycle is $(m-1)^2$. Let θ_1 be the $(m+1)^{\text{th}}$ term in the cycle (75), θ_2 the $(2m+1)^{\text{th}}$ term, and so on. Then the root of an equation of the mth degree, including the instances above given, is

$$r_1 = (\theta + \theta^{-1}) + (\theta_1 + \theta_1^{-1}) + \ldots + (\theta_{\frac{m-3}{2}} + \theta_{\frac{m-3}{2}}^{-1}).$$

www.ingramcontent.com/pod-product-compliance
Lightning Source LLC
Chambersburg PA
CBHW031816090426
42739CB00008B/1292